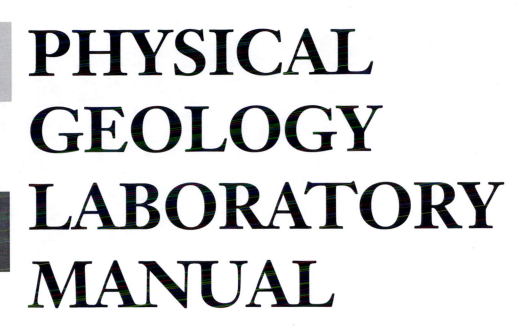

PHYSICAL GEOLOGY LABORATORY MANUAL

Fifth Edition
Revised Printing

Karen M. Woods
Lamar University

Contributing Authors

Margaret S. Stevens
James B. Stevens
Roger W. Cooper
Donald E. Owen
James Westgate
Jim L. Jordan
Bennetta Schmidt

Photography

Doug Sampson

Kendall Hunt
publishing company

Cover images provided by the author.

Kendall Hunt
publishing company

www.kendallhunt.com
Send all inquiries to:
4050 Westmark Drive
Dubuque, IA 52004-1840

Copyright © 1994, 1997, 2001, 2006, 2012 by Kendall Hunt Publishing Company
Revised Printing 2013.

ISBN 978-1-4652-2313-5

Printed in the United States of America
10 9 8 7 6 5

CONTENTS

PREFACE

Physical Geology is the first introductory course in the field of Geology. The faculty and staff of Lamar University, Department of Earth and Space Sciences have collaborated to produce a laboratory manual that is informative and easily understood. It has been customized to present the concepts and ideas the faculty feel are most important in Physical Geology. It is intended to supplement the main lecture course by exposing the student to conceptual exercises and hands-on experience of the subjects introduced in lecture.

MINERAL **PRELAB** WORKSHEET

Name: _____   Sect.: _____

What are the five characteristics required for a substance to be considered a mineral?

1. _____

2. _____

3. _____

4. _____

5. _____

Which characteristic of minerals is **not** a characteristic of mineraloids? _____

Sodium (an explosive metal) and chlorine (a poisonous gas) combine with one another (NaCl) to form a mineral that is an important requirement for a healthy life. This mineral can also cause health problems if excessive amounts are included in the diet.

What is the **name of this mineral**? _____

Define crystal form. _____

_____

_____

_____

Define cleavage. _____

_____

_____

_____

How is crystal form different from cleavage? (Explain completely.)

_____

_____

_____

_____

What type of fracture is described as smoothly curved? _____

Do all minerals have cleavage? _____

Can minerals have both cleavage and fracture? _____

What is the difference between cleavage and fracture? (Explain completely.)

_____

_____

_____

Which physical property of minerals is described as the **least useful as the sole identifying property** of minerals? _____

What mineral has a hardness of 2 on Mohs' scale? _____

What mineral has a hardness of 4 on Mohs' scale? _____

What mineral has a hardness of 9 on Mohs' scale? _____

Do all minerals have an exact hardness value? _____

What are parallel grooves seen on either cleavage planes or crystal faces?_____

Does orthoclase feldspar have striations? _____

List four minerals that may exhibit striations.

_____   _____

_____   _____

What is streak? _____

_____

What are the two major divisions that describe luster? _____   _____

List four minerals with a shiny metallic luster.

_____   _____

_____   _____

Describe the nonmetallic luster of the following minerals. Quartz _____

Muscovite _____   Malachite _____

What crystal form do the following minerals have, if visible?

Quartz _____   Galena and Halite _____

Garnet _____   Fluorite _____

Pyrite (two possible) _____ and _____

How many planes of cleavage do the following minerals have, when present?

Halite _____   Orthoclase _____   Fluorite _____   Calcite _____

What are the three terms used to describe diaphaneity? _____

_____   _____

Which mineral(s) react(s) **vigorously** to hydrochloric acid?

_____

Heavy, light, and *normal* are terms that describe which physical property? _____

Minerals

C H A P T E R 1

INTRODUCTION

Geology deals with the physical and historical aspects of the Earth. **Physical geology** is the study of the composition, behavior, and processes that shape the Earth. The science of geology also provides the means to discover and utilize the Earth's natural resources (coal, natural gas, petroleum, minerals, etc.). Geologists also study the Earth and its processes to better understand and predict potentially dangerous geologic situations (earthquakes, volcanic eruptions, floods, etc.) in order to save lives. **Historical geology** is the study of the geological history of the Earth.

MINERALS

Minerals are the basic building blocks of nearly all of Earth's materials. A **mineral** is a naturally occurring, solid, inorganic compound of one or more elements, has an orderly arrangement of atoms (crystallinity), and has an established chemical composition that can vary slightly within specific limits. A material that satisfies these requirements will have a characteristic set of physical properties that can be used for identification.

Natural compounds are not "pure" in the pharmaceutical sense, particularly if modern analytic methods are used. Most chemical elements can be shown to consist of several isotopes (atoms of different atomic weights that have a closely similar set of chemical properties). Minerals as natural compounds are fairly complicated. They consist of one or more elements, and each element consists of one or more isotopes. They aren't absolutely "pure" compounds and show some variation, even within materials called by the same mineral name. The guideline geologists have agreed on to define a particular mineral is the nature of the internal geometric arrangement (the crystallinity) of the atoms. This arrangement is usually called the crystal structure. Materials such as glass, opal, and bauxite (an important ore of aluminum) have no particular geometric arrangement of their atoms, and are not true minerals because they lack crystallinity. These materials are referred to as **mineraloids**.

SUMMARY: A material must have the following characteristics to be classified as a mineral:

1. be naturally occurring (not man-made),
2. be solid,

3. be inorganic (not compounds that can only be produced by living organisms),

4. have a geometric arrangement of its atoms—crystallinity, and

5. have a chemical composition that can vary only according to specific limits.

A substance that satisfies those requirements will have a characteristic set of physical properties that can be used for identification.

COMMON MINERALS

Many of the minerals studied in introductory geology (Table 1.1) are familiar to both geologists and non-geologists alike. The easily recognizable native minerals include sulfur, graphite, and diamond. These are considered minerals when found in large, natural, cohesive quantities. The most commonly known mineral is quartz (SiO_2, silicon dioxide). Varieties include: rose quartz (pink), milky quartz (white-gray), rock crystal quartz (clear), amethyst (purple), aventurine (green), and others. Halite, Sodium Chloride, (NaCl), is table salt, found in almost every spice cabinet. Minerals have many diverse and unexpected uses, some of which are included in a list at the end of the chapter.

PHYSICAL PROPERTIES

All minerals have a distinctive set of physical properties that can be used to identify them when looking at hand samples. The goal of the student is to learn the geological terminology and processes and apply them to unknown mineral specimens.

Students should note that the **physical properties** of minerals <u>are *not absolutes*</u>. Hardness is a physical property of minerals that can vary slightly between different samples of the same mineral. Magnetite, for example, has a hardness that ranges from 5.0–6.0 on the hardness scale. This means that some samples will easily scratch a glass plate (hardness of 5.5) and some will not be able to scratch it at all. Color is another property of minerals that varies widely within mineral samples and therefore should not be the only criteria used to identify minerals. Quartz is a mineral that is found in many different colors, for example, as mentioned previously. The students should not use any one property alone to identify mineral unknowns. A group of physical properties leads to a more accurate identification.

Crystal Form

Crystal form is the geometric arrangement of plane (flat) surfaces on the outside of a mineral that reflects the internal crystallinity of the mineral. *Crystal faces* are the planar (flat) sides of a cube or a dodecahedron, for example, and only develop when the mineral has the time and space to grow without interference. Halite and fluorite often develop **cubic crystal form**. A cube has six planar surfaces. The mineral garnet can develop **dodecahedral crystal form,** and pyrite develops either **cubic** or **pyritohedral crystal form**. Corundum, quartz, and apatite show different variations on the **hexagonal (six-sided) crystal form**. Calcite can develop hexagonal crystal form (difficult to recognize) or **scalenohedral crystal form**. Figure 1.1 illustrates common crystal forms and associated minerals. Minerals without an external crystal form are referred to as **massive**. Massive minerals include magnetite, hematite, talc, some quartz, and others. Crystals grow as "invisible atoms" of a solution bond together in a given geometric framework that is consistent with the atoms' electrical or size characteristics. As the geometric framework enlarges with continued "growth," the geometry becomes visible as smooth surfaces (crystal faces). These faces give crystals of various minerals their characteristic shape and beauty.

TABLE 1.1   Chemical Groups of Selected Minerals

Chemical Class	Mineral	Chemical Composition	
Natives Only one element present, "naturally pure"	Sulfur Graphite/Diamond	S C	(Sulfur) (Carbon)
Oxides A metal bonds directly with an oxygen as the nonmetal	Quartz (All varieties) Hematite (All varieties) Goethite Magnetite Corundum Topaz Bauxite (Mineraloid)	SiO_2 Fe_2O_3 $FeO(OH)$ Fe_3O_4 Al_2O_3 $Al_2SiO_4(OH,F)_2$ $Al_2O_3nH_2O$	(Silicon dioxide) (Iron oxide) (Hydrous iron oxide) (Iron oxide) (Aluminum oxide) (Aluminum silicate hydroxide) (Hydrous aluminum oxide)
Sulfides A metal bonds directly with sulfur as the nonmetal	Pyrite Galena Sphalerite	FeS_2 PbS ZnS	(Iron sulfide) (Lead sulfide) (Zinc sulfide)
Sulfates A metal bonds with the SO_4 complex ion acting as a nonmetal	Gypsum (All varieties) Anhydrite	$CaSO_4 . 2H_2O$ $CaSO_4$	(Hydrous calcium sulfate) (Calcium sulfate)
Carbonates A metal bonds with the CO_3 complex ion acting as a nonmetal	Calcite Dolomite Malachite	$CaCO_3$ $MgCaCO_3$ Cu_2CO_3	(Calcium carbonate) (Magnesium-calcium carbonate) (Copper Carbonate)
Halides A metal bonds with a halogen (Cl, F, Br or I) as the nonmetal	Halite Fluorite	$NaCl$ CaF_2	(Sodium chloride) (Calcium fluoride)
Phosphates A metal bonds with the PO_4 complex ion as the nonmetal	Apatite	$Ca_5(PO_4)_3(F,Cl,OH)$ (Calcium phosphate)	
Silicates A metal bonds with the SiO_4 complex ion acting as the nonmetal			
Nesosilicates (Island silicates)	Olivine Garnet	$(Fe,Mg)SiO_4$ (Iron-magnesium silicate) Complex Ca, Mg, Fe, Al silicate	
Inosilicates (Chain silicates)	Hornblende (Amphibolite) Augite	Ca, Na, Fe, Mg, Al silicate $(Ca,Na)(Mg,Fe,Al)(Si,Al)_2O_6$	
Phyllosilicates (Sheet silicates)	Muscovite Biotite Chlorite Talc Kaolinite	$KAl_2(AlSi_3O_{10})(OH)_2$ See Mineral Properties List See Mineral Properties List $Mg_3Si_4O_{10}(OH)_2$ $Al_4Si_4O_{10}(OH)_8$	
Tectosilicates (3-D silicates)	Orthoclase feldspar Plagioclase feldspar (Labradorite) Quartz	$KAlSi_3O_8$ (Potassium-aluminum silicate) $(Ca,Na)AlSi_3O_8$ (Calcium, sodium-aluminum silicate) SiO_2 (Silicon dioxide)	

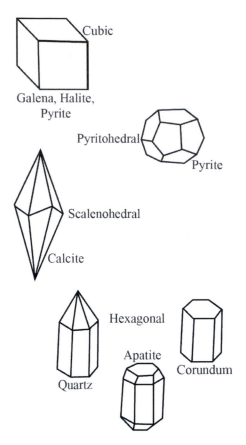

FIGURE 1.1 Crystal Form.

Crystal Habit

Crystal habit is the general shape a mineral develops. Although many minerals develop well-defined crystal forms, most do not, but they do tend to develop in specific shapes or **habits**. Table 1.2 is a list of common crystal habits and associated minerals. *Habits indicated by an asterisk are also crystal forms.*

Cleavage

Cleavage is the tendency of a mineral to *break* in a **systematic** (regular or ordered) way, along planes of weakness determined by the strength of chemical bonds between the atoms that make up the mineral. The cleavages (planes of weakness) represent layers between rows or sets of planar atoms where the atomic bonds are weakest. See the lecture text for an explanation of the bonds between atoms. Some minerals, such as gypsum and micas, have one plane of cleavage, but most have multiple planes of cleavage. Figure 1.2 illustrates six different cleavage patterns and associated minerals. *Although cleavage is a very useful identifying feature when present, not all specimens of a given mineral will have visible or easily identifiable cleavage planes.* Some cleavages are microscopic, invisible to the naked eye, and therefore not useful for identification of mineral unknowns. Cleavage is classified as *perfect* (readily recognizable), *good, fair,* and *poor* (difficult to distinguish). Even when cleavage is not visible on a particular hand specimen, it does not mean that the mineral lacks cleavage. Look at multiple specimens of the same mineral before determining the presence or absence of cleavage. Because many minerals do not have cleavage or have

TABLE 1.2   Crystal Habit

Crystal Habit/Crystal Form	Associated Minerals
*Cubic	Fluorite, Halite, Pyrite
*Dodecahedral	Garnet
*Octahedral	Fluorite
Prismatic	Apatite
*Pyritohedral	Pyrite
*Rhombohedral	Calcite
*Scalenohedral	Calcite
Columnar	Hornblende
Fibrous	Satin spar gypsum
Granular	Olivine, Alabaster gypsum
Oolitic	Hematite
Micaceous	Muscovite, Biotite
Radiating	Goethite
Hexagonal	Quartz, Apatite, Corundum

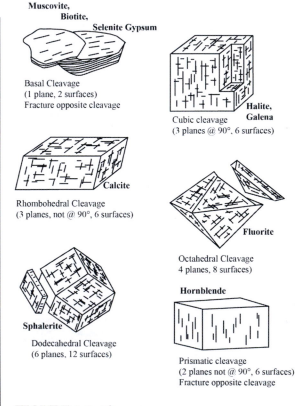

FIGURE 1.2 Cleavage.

microscopic cleavage, the presence of cleavage can eliminate all minerals that do not have cleavage. Some minerals always have easily identifiable cleavage. Muscovite and biotite (micas) both have perfect cleavage in one direction. They cleave (split) into thin, flat, flexible sheets. The two parallel flat sides, the top and bottom, represent one cleavage plane. Two planes of cleavage will have two sets of parallel flat sides. In other words, you can count the number of flats sides and divide this by two to determine the number of planes of cleavage.

Unfortunately, cleavage and crystal form are easily confused. They both result in flat planar surfaces, but for different reasons. The planar faces of crystal form develop as the result of the growth of a mineral into a geometric shape, and cleavage planes are the result of breakage along planes of weakness. *Minerals with cleavage will break in the same direction or set of directions every time and form flat planar surfaces or a stair-step pattern along the cleavage planes.* Some minerals have both cleavage and crystal form (halite, fluorite, etc.), some only have crystal form (quartz), and some only have cleavage (muscovite). Minerals with crystal form will break in no particular direction and develop irregular (uneven) surfaces when broken.

TABLE 1.3   Fracture

Fracture	Characteristics	Associated Minerals
Conchoidal	Smoothly curved	Quartz, alabaster gypsum
Uneven	Rough and irregular	Magnetite, hematite
Splintery	Long, thin fragments	Hornblende

Fracture

Fracture is the nonsystematic and irregular way some minerals break. The fracture surface is rough or uneven, unlike cleavage planes, which are flat and often smooth. **Conchoidal fracture** is a special kind of breakage that results in a curved parting surface. When a bullet passes through glass, a curved or listric surface is produced. Conchoidal fracture occurs within homogeneous materials. Homogeneous materials, such as glass, lack planes of weakness—the material is equally strong in all directions. Quartz commonly shows conchoidal fracture. Many minerals have both cleavage and fracture, such as plagioclase feldspar (labradorite). Labradorite has two directions of cleavage and fracture opposite the cleavage directions. Table 1.3 lists fracture types, characteristics, and associated minerals.

Striations

Striations are very fine, parallel grooves visible on cleavage planes or crystal faces of some minerals due to their crystal structure and growth patterns. Plagioclase feldspar (labradorite and albite) commonly has striations on one cleavage plane. The striations on plagioclase become increasingly obvious as the calcium content of the feldspar increases. Striations can also be visible on the crystal faces of minerals such as pyrite, quartz, and corundum. Striations become more visible when the mineral is slightly rotated back and forth in the light. The striations will reflect the light as the mineral is turned. As with cleavage, striations *may or may not be visible* on individual samples of the same mineral. Quite often, the striations are not visible on some plagioclase feldspars, for example, and on others they are very obvious. Although the absence of striations on a mineral cannot be used to identify specific minerals, the presence of striations can be used to eliminate those minerals that do not have striations as a physical characteristic.

Tenacity

Tenacity is the resistance of a mineral to breakage. Many minerals are easily broken, whereas some are very hard to break. Terms used to describe tenacity include *brittle, elastic, malleable,* and so forth. Gold, a soft mineral, is malleable and easily deformed when hit. Diamond, the hardest known natural mineral, is brittle and will shatter easily when hit. **DO NOT TEST** the tenacity of the mineral specimens given unless told to do so by the instructor.

TABLE 1.4 Mohs' Scale of Hardness

Hardness	Mineral	Common Testing Tools
10	Diamond	
9	Corundum	
8	Topaz	
7	Quartz	
6	Orthoclase	
5.5		Glass plate, steel knife, nail
5	Apatite	
4	Fluorite	
3.5		Copper penny
3	Calcite	
2.5		Natural fingernail
2	Gypsum	
1	Talc	

Hardness

Hardness is a mineral's resistance to being scratched. Some minerals are very soft and easily scratched by a natural fingernail, whereas some are hard enough to scratch a glass plate. The relative hardness of a mineral can be determined by using **Mohs' Scale of Hardness** (Table 1.4). The German mineralogist Friedrich Mohs (1773–1839) developed the hardness scale. Mohs arranged ten common or certain unique minerals in order of their increasing relative hardness to provide a standard scale to which all minerals can be compared. Talc was chosen by Mohs to represent the hardness of 1 (softest) and diamond to represent 10 (hardest). Some common everyday materials also fit conveniently into Mohs' scale as well and serve to determine a more accurate hardness range for unknown minerals. These include: natural fingernails (hardness = 2.5), copper penny (hardness = 3.5), and steel nail, knife, and glass plate (hardness 5.5).

The best way to determine the hardness of a mineral unknown is to find the softest material that will scratch it. For example, an unknown mineral that cannot be scratched by a fingernail but can be scratched by a penny has a hardness that is greater than 2.5 but less than 3.5. The range recorded would be 2.5–3.5. For minerals not on Mohs' scale, the student should record the smallest hardness range possible. If the mineral is on Mohs' scale, the exact hardness should be used. Memorization of the hardness of all minerals is not necessary, although the minerals on Mohs' scale should be.

In general, minerals can be separated into two groups, those that are greater than 5.5 (will scratch glass) and those that are less than 5.5 (cannot scratch glass). These can be further refined into those that are between 3.5–5.5 (harder than a penny but less than glass), between 2.5–3.5 (less than a penny but harder than a fingernail), or less than 2.5 (can be scratched by a fingernail). Hardness alone should not be used to identify mineral unknowns.

Color

Color is a function of how the surface of a mineral reflects or absorbs white light. Color is one of the *least useful properties* of minerals because very few have a consistent color. Exceptions include: sulfur (bright yellow), pyrite (brassy yellow), and galena (gray). Quartz and calcite are examples of minerals with a wide array of color variations. They can be green, yellow, black, clear, and so forth.

There are three general causes of color variation in minerals:

1. Impurities within the mineral.
2. The disturbance of the crystallinity of the mineral.
3. The size of the mineral pieces. Thin pieces are usually lighter in color than thicker pieces (one of the most common causes of color variation).

Although color alone cannot be used to identify mineral unknowns, minerals can be sorted into groups of darker and lighter hues before testing other physical properties.

Streak

Streak is the color of a mineral's powder (or the color of the mineral when the crystals are very small). The streak is obtained by rubbing the mineral on an unglazed ceramic or porcelain plate.

Gently blow or shake off as much of the powder as possible off the plate. *The color of the powder that sticks to the streak plate is the actual streak.* Most minerals will have a white or clear streak, but many have a colored streak. Hematite is a mineral with a red to reddish-brown streak. Hematite can be the same color as the streak (red to reddish-brown) or it can be a silvery metallic color. The color of the mineral does not determine the color of the resulting streak.

Luster

Luster is the way a mineral reflects light. Luster is described either as *metallic* (fresh un-tarnished metal) or *nonmetallic*. Types of nonmetallic luster include pearly, greasy, silky, vit-reous (glassy), earthy, waxy, resinous, and so forth.

Magnetism

Magnetism is the attraction of a magnet to a mineral. Minerals range from not magnetic (garnet) to slightly magnetic (some hematite) to strongly magnetic (magnetite).

Specific Gravity

Density is mass per unit volume. **Specific gravity** is the ratio of the density of a given ma-terial to the density of an equal volume of water (at 4°C). Minerals that feel unusually heavy for their size have a high specific gravity. Galena is one example. Some minerals feel lighter than their size would indicate, and other minerals feel normal weight for their size. In most instances, specific gravity cannot be used for mineral identification, but if a min-eral is unusually heavy or lightweight, it is a useful physical property.

Diaphaneity

Diaphaneity refers to how and what extent light is transmitted through a mineral. A thin-section is a 0.03-mm slice of a mineral that is thin enough to allow light to pass through it. Although diaphaneity is usually applied to thin-sections, we will apply the same terminology to the hand samples seen in lab. The diaphaneity can be determined by looking at the minerals.

1. **Transparent:** Light passes easily through the mineral; thus, images can be clearly seen through it. Rock crystal quartz is transparent.
2. **Translucent:** Some light passes through the mineral, but light is diffused and ab-sorbed internally by the mineral; thus, the image cannot be seen clearly. Translucency is a matter of the thickness and purity of the mineral. Hematite is usually thought of as opaque, but extremely small, pure crystals are translucent. Although pure quartz is clear and colorless, the presence of large numbers of small bubbles can make it translucent (milky quartz). Disturbance of the crystal by radiation and from decaying radioactive elements can make quartz gray, brown, or black, and the crystal, particu-larly if thick, may be translucent, or nearly opaque.
3. **Opaque:** The mineral allows no light to pass through; thus, images cannot be seen through the mineral. The thickness and purity of a mineral affect the opacity. Metallic or submetallic minerals (pyrite, magnetite) are opaque even as thin sections.

Double Refraction

Double refraction is the doubling of a single image seen through a transparent mineral. Minerals, except the cubic ones such as fluorite, halite, and diamond, split light rays into two parts that follow different paths as they pass through the crystal. Optical-quality calcite crystals are the best example of this because the two parts of the light follow very different paths. To see double refraction, place an example of optical-quality calcite on this page and look at the words. Special microscopes and specially prepared specimens are used in serious work with double refraction, but geologists frequently make use of this property in hand specimen mineral identification.

Reaction to Hydrochloric Acid (HCl)

Some minerals will chemically react (fizz, or give off H_2O and bubbles of CO_2) with dilute hydrochloric acid (HCl). This test is primarily used to identify calcite ($CaCO_3$) and dolomite [$CaMg(CO_3)_2$]. Calcite reacts strongly with cool, dilute HCl, and most dolomites only react when powdered. Scratch dolomite with a nail to produce enough powder to test its reaction with acid. Apply one or two drops directly to the calcite or the dolomite powder and note the reaction. *After the acid is applied and the result noted, wipe the excess acid off the mineral with a paper towel.*

CAUTION: All students are to wear safety goggles when using acid.

Other Identifying Properties

Smell, touch, and *taste* can also be used to identify mineral unknowns. Many minerals have strong smells, such as sulfur and sphalerite (like rotten eggs). Like a child's "scratch-and-sniff" book, scratch the minerals with a nail and sniff. Some minerals have a particular feel to them. Graphite (pencil lead) and halite feel greasy, whereas kaolinite feels powdery. The last property is taste. Halite, common table salt, tastes salty. **DO NOT TASTE ANY MINERALS IN LAB**. When using taste as an identifying characteristic, a clean, fresh surface is advisable. *The minerals used in lab are not safe to taste.*

A **Mineral Properties List** and a **Mineral Uses List** have been included for student use. The Mineral Properties List describes the properties of ideal specimens. Student lab samples, in general, are not "ideal" specimens and will not perfectly display every property listed. The Mineral Properties List should be used as an aid to confirm the identification of the mineral unknowns. The Mineral Uses List describes one or more possible uses for each mineral.

> Mineral pictures and other information can be found on the Earth and Space Science website (http://ess.lamar.edu). Click on the People tab, Staff, Karen M. Woods, Teaching, Physical Geology Lab, Minerals.

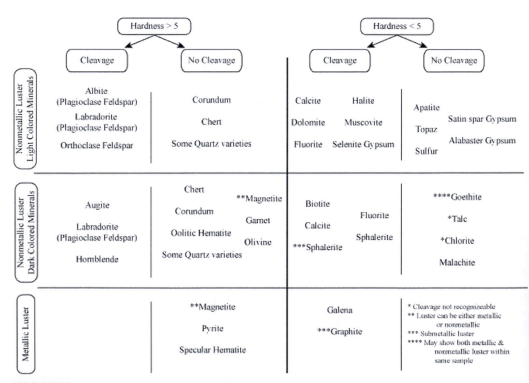

FIGURE 1.3 Mineral Identification Key.

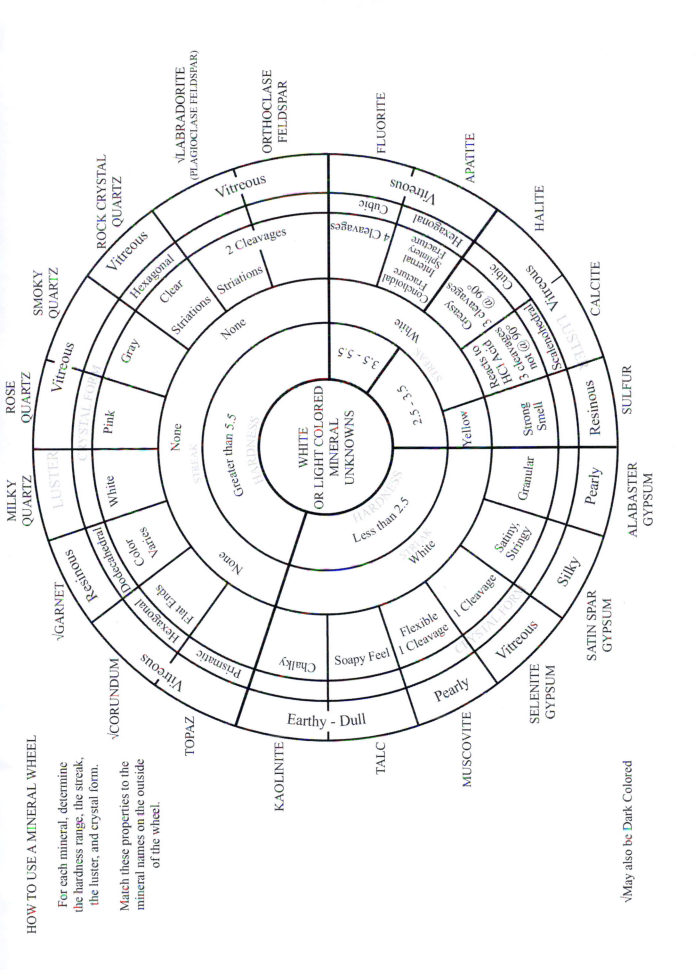

HOW TO USE A MINERAL WHEEL

For each mineral, determine the hardness range, the streak, the luster, and crystal form.

Match these properties to the mineral names on the outside of the wheel.

√May also be Dark Colored

WHITE OR LIGHT COLORED MINERAL UNKNOWNS

HARDNESS
Greater than 5.5
Less than 2.5
2.5 - 3.5
3.5 - 5.5

STREAK
None
White
Yellow
White

CRYSTAL FORM
Hexagonal
Clear
Gray
Pink
White
Dodecahedral
Color Varies
Hexagonal Flat Ends
Prismatic
Cubic
Hexagonal
Scalenohedral
Cubic
Granular
Satiny, Stringy
1 Cleavage
Flexible 1 Cleavage
Soapy Feel
Chalky

LUSTER
Vitreous
Vitreous
Vitreous
Vitreous
Resinous
Vitreous
Vitreous
Strong Smell
Resinous
Pearly
Silky
Vitreous
Pearly
Earthy - Dull

STRIATIONS
2 Cleavages
Striations
Striations
None
None
4 Cleavages
Internal Spintery Fracture
Conchoidal Fracture
Greasy 3 cleavages not @ 90°
3 cleavages @ 90°
Reacts to HCl Acid

ROCK CRYSTAL QUARTZ
SMOKY QUARTZ
ROSE QUARTZ
MILKY QUARTZ
√GARNET
√CORUNDUM
TOPAZ
KAOLINITE
TALC
MUSCOVITE
SELENITE GYPSUM
SATIN SPAR GYPSUM
ALABASTER GYPSUM
SULFUR
CALCITE
HALITE
APATITE
FLUORITE
ORTHOCLASE FELDSPAR
√LABRADORITE (PLAGIOCLASE FELDSPAR)

11

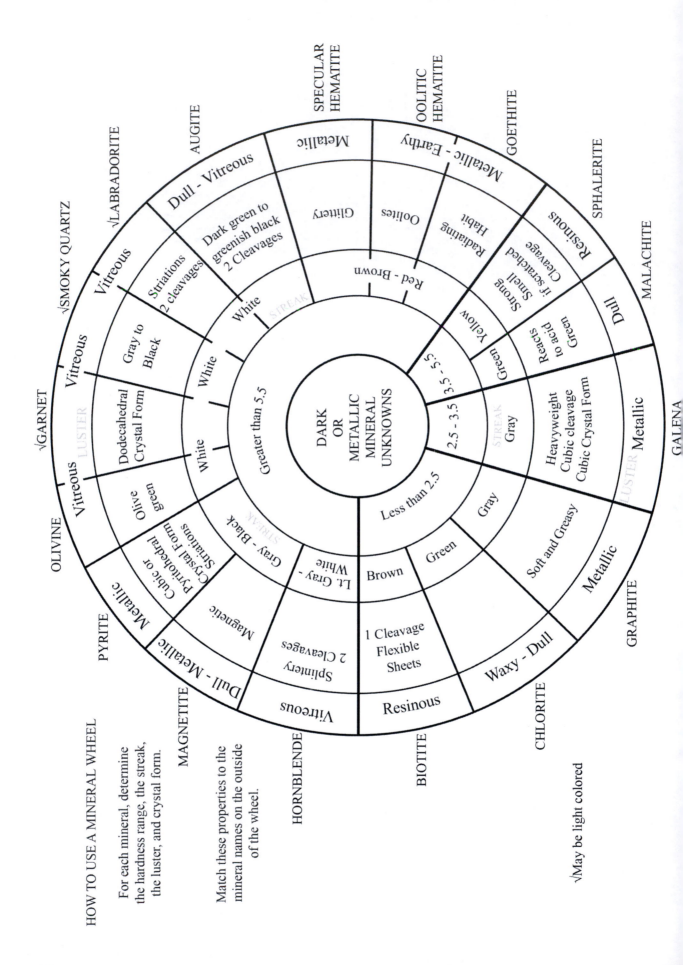

HOW TO USE A MINERAL WHEEL

For each mineral, determine the hardness range, the streak, the luster, and crystal form.

Match these properties to the mineral names on the outside of the wheel.

√May be light colored

MINERAL PROPERTY LIST

*Albite—Albite is a low-temperature, light-colored, plagioclase feldspar with two directions of cleavage and two opposing sides with uneven fracture. Striations may be present. Albite has a vitreous luster, a specific gravity of 2.62–2.76, and a hardness of 6. Chemical formula: $NaAlSi_3O_6$ (sodium aluminum silicate).

Apatite—Apatite is usually a shade of green or brown, and has a vitreous luster, a specific gravity of 3.1–3.2, a white streak, conchoidal or uneven fracture, and a hardness of 5 on Mohs' scale. Apatite has hexagonal crystal form and looks cracked throughout the crystal. Chemical formula: $Ca_5(PO_4)_3(F,Cl,OH)$ (calcium phosphate).

Augite—Augite is a pyroxene with two cleavage planes, one at 87° and the other at 93°. Augite is dark green to black, and has a vitreous to dull luster, a specific gravity of 3–3.5, and a hardness that ranges from 5 to 6; it lacks a streak. Other identifiable properties include a hackly or splintery fracture opposite to the cleavage direction. Chemical formula: $(Ca,Na)(Mg,Fe,Al)(Si,Al)_2O_6$ (calcium, sodium, magnesium, iron, aluminum silicate).

*Bauxite—Bauxite is a mineraloid. Bauxite is brown, gray, white, or yellow, and has a dull to earthy luster, no cleavage, a white to yellow-brown streak, and a hardness that ranges from 1 to 3. Bauxite usually occurs in compact masses of pisoliths (pea-sized concretions, spheres coarser than ooliths). Fracture is uneven. Chemical formula: $AlO(OH)$ (hydrous aluminum oxide).

Biotite—Biotite is a black to dark-brown mineral with a vitreous to pearly luster. Biotite has perfect cleavage in one direction, allowing it to be separated into thin sheets. Biotite has a brown to dark-green streak if the specimen is big enough, and a hardness of 2.5–3. Fracture is uneven perpendicular to cleavage direction. Chemical formula: $K(Mg,Fe)_3(AlSi_3O_{10})(OH)_2$ (hydrous potassium, magnesium, iron, aluminum silicate).

Calcite—Calcite is usually white to colorless, but may be yellow, green, blue, red, black, or other colors due to impurities. Calcite has perfect rhombohedral cleavage (see photo), hexagonal crystal form (if present), a white to gray streak, and a vitreous to earthy luster. Hardness is 3 on Mohs' scale. Specific gravity is 2.71. Calcite is soluble in dilute hydrochloric acid with a strong effervescence (fizz). Double refraction is visible through colorless rhombs. Chemical formula: $CaCO_3$ (calcium carbonate).

Chlorite—Chlorite is a green to greenish-black mineral with a waxy to earthy luster. Chlorite has a perfect basal cleavage (not apparent in massive pieces), and its streak is pale green to white. The specific gravity is 3 and hardness is 2–2.5. Chlorite feels slippery. Chemical formula: $(Mg,Fe)_3(Si,Al)_4O_{10}(OH)_2 \cdot (Mg,Fe)_3(OH)_6$ (magnesium, iron, aluminum silicate).

Corundum—Corundum varies in color (brown, blue, red, etc.), has an adamantine to vitreous luster, has a hardness of 9 on Mohs' scale, and has a specific gravity of 4. Corundum is found in massive deposits as emery and as hexagonal crystals (see photo) with striations on basal faces and has conchoidal fracture. Gem-quality corundum is commonly known as sapphire and ruby. Chemical formula: Al_2O_3 (aluminum oxide).

Dolomite—Dolomite varies from colorless to white, gray, brown, and pink (crystals). Dolomite has perfect rhombohedral cleavage, hexagonal crystal form, and a vitreous to pearly luster. Cleavage and crystal form are not evident in massive pieces. Specific gravity is 2.85, hardness is 3.5–4, and dolomite has a white streak. In powdered form, dolomite effervesces in cold, dilute hydrochloric acid. Chemical formula: $CaMg(CO_3)_2$ (calcium, magnesium carbonate).

Fluorite—Fluorite has perfect octahedral cleavage, cubic crystal form, and conchoidal fracture. Fluorite is colorless and transparent when pure but may be blue, green, pink, purple, yellow, or black. Fluorite has a vitreous luster, specific gravity of 3.18, hardness of 4, and a white streak. Chemical formula: CaF_2 (calcium fluoride).

Galena—Galena has a perfect cubic cleavage and cubic or octahedral crystal form. Galena is lead gray, and has a gray streak, metallic luster, and a hardness of 2.5. Galena has a high specific gravity (7.57). Chemical formula: PbS (lead sulfide).

Garnet—Garnet has a splintery or conchoidal fracture, no cleavage, and a resinous to vitreous luster. Color varies with composition but is commonly dark red to reddish-brown or yellow. Garnet forms dodecahedral crystals in some metamorphic rocks and is also found in coarse granular masses. Garnet has a specific gravity of 3.5–4.3, and a hardness of 6.5–7.5. Chemical formula: Fe, Mg, Mn, Ca, Al silicate (complex iron, magnesium, manganese, calcium, aluminum silicate).

Goethite—Goethite is a variety of iron oxide. Goethite has a prismatic crystal form and cleaves parallel with the prisms. Goethite is yellow or yellowish brown to silvery brown in color, and has a brownish-yellow streak, a specific gravity of 4.37, and a hardness that ranges from 5 to 5.5. Massive goethite has an adamantine to dull luster. Goethite is also found with rounded (reniform) masses that have a metallic luster. Chemical formula: FeO(OH) (hydrous iron oxide). Pronounced "guhr-thite."

Graphite—Graphite has perfect cleavage in one direction, although the mineral is usually found as foliated masses. Graphite is dark gray to black in color, and has a gray to black streak, metallic luster, a specific gravity of 2.23 (low), and a hardness of 1–2. Graphite feels "greasy." Chemical formula: C (carbon).

Gypsum—Gypsum is translucent and generally white, but may be tinted to various colors. Gypsum has a white streak; pearly to vitreous luster; cleavage; a conchoidal, irregular, or fibrous fracture; a specific gravity of 2.32; and a hardness of 2 on Mohs' scale. Chemical formula: $CaSO_4 \cdot 2H_2O$ (hydrous calcium sulfate). Three varieties are distinctive.

> **Alabaster gypsum**—Alabaster is the fine-grained, massive variety of gypsum. Alabaster, also called rock gypsum, is generally white, but may be slightly tinted with other colors. It has a pearly luster and cleavage is not apparent. Chemical formula: See above.

> **Selenite gypsum**—Selenite gypsum has perfect cleavage in one direction, and a conchoidal fracture. Selenite is colorless to white, transparent to translucent, and has a vitreous luster. Chemical formula: See above.

> **Satin spar gypsum**—Satin spar gypsum is fibrous, colorless to white, and has a silky luster. Cleavage is not apparent in this variety. Chemical formula: See above.

Halite—Halite has perfect cubic cleavage and cubic crystal form (see photo). Halite is colorless to white but impurities can give it a yellow, red, blue, or purple tint. Halite is transparent to translucent, and has a vitreous luster, a specific gravity of 2.16, and a hardness of 2.5. Halite feels greasy and tastes salty (tasting of laboratory specimens is not recommended). Chemical formula: NaCl (sodium chloride).

Hematite—Hematite is steel gray, to black, to red, to reddish brown. Hematite has a red to red-brown streak, a specific gravity of 5.26, a hardness that ranges from 5.5 to 6.5, an irregular fracture, and a metallic or dull luster. Chemical formula: Fe_2O_3 (iron oxide). Oolitic and specular are two important varieties.

> **Oolitic hematite**—Oolitic hematite is composed of small spheres (ooliths) of hematite. Oolitic hematite is red to brownish red, has a red streak, and has an earthy luster. See hematite above for other properties. Chemical formula: See above.

> **Specular hematite**—Specular hematite has a platy (glitter-like) appearance and may be slightly to strongly magnetic. Specular hematite is steel gray or "silvery" with a metallic luster, and has a red streak. See hematite above for other properties. Chemical formula: See above.

Hornblende—Hornblende is dark green to black, and has a vitreous luster, a specific gravity of 3–3.5, a white to gray streak, and a hardness of 5 to 6. Hornblende is an amphibole with two cleavage angles (56° and 124°) and an uneven fracture opposite of the cleavage directions. Chemical formula: Ca, Na, Mg, Fe, Al silicate (calcium, sodium, magnesium, iron, aluminum silicate).

Kaolinite—Kaolinite has perfect cleavage (not apparent in massive pieces). Kaolinite is white, and has a dull to earthy luster, a white streak, a specific gravity of 2.6, and a hardness of 2. Kaolinite looks and feels like chalk, a kind of limestone, but does not react with hydrochloric acid. Kaolinite fractures irregularly. Chemical formula: $Al_4Si_4O_{10}(OH)_8$ (hydrous aluminum silicate).

Labradorite—Labradorite is gray-blue, medium-temperature, plagioclase feldspar with two directions of cleavage, and two opposing sides with uneven fracture. Some samples exhibit a flash ("play") of different colors on cleavage surfaces. Striations may be present. Labradorite has a vitreous luster, a specific gravity of 2.62 to 2.76, and a hardness of 6. Chemical formula: $(Ca,Na)AlSi_3O_8$ (calcium-sodium aluminum silicate).

*Limonite—Limonite, a variety of iron oxide, is dark brown to orange to brownish yellow, and has a yellow to brown streak, an earthy to dull luster, a specific gravity of 2.9–4.3, and a hardness of 4–5.5. Limonite fractures irregularly. Chemical formula: $FeO(OH)$ (hydrous iron oxide).

Magnetite—Magnetite is a black mineral with a gray to black streak, a specific gravity of 5, a hardness of 5.5–6, and a dull luster. It is strongly magnetic, and fractures irregularly. Chemical formula: Fe_3O_4 (iron oxide).

Malachite—Malachite is a bright-green mineral with a pale-green streak, a specific gravity of 4.05, a hardness of 3.5–4, and a silky to dull luster; it fractures irregularly. Malachite is frequently reniform (rounded, kidney-shaped masses). Reacts strongly to HCl acid. Chemical formula: $Cu_2CO_3(OH)_2$ (copper carbonate).

Muscovite—Muscovite is colorless to brown, gray, or green. Muscovite has a vitreous to silky to pearly luster; perfect cleavage in one direction, allowing it to be separated into thin flexible sheets; a white streak (if sample is thick enough); a specific gravity of 2.8; and a hardness of 2–2.5. Fracture is uneven perpendicular to the cleavage direction. Crystal system: monoclinic. Chemical formula: $KAl_2(AlSi_3O_{10})(OH)_2$ (hydrous potassium, aluminum silicate).

Olivine—Olivine is an olive-green to light-gray mineral with a vitreous luster, conchoidal fracture, a specific gravity of 3, and a hardness of 6.5–7. Cleavage, when visible, is poor. Chemical formula: $(Mg,Fe)_2SiO_4$ (magnesium, iron silicate).

Orthoclase Feldspar—Orthoclase feldspar is white to pink, and has a vitreous luster, a specific gravity of 2.57, and hardness of 6 on Mohs' scale. Orthoclase has two directions of cleavage at 90° and an uneven fracture opposite the cleavage directions. Chemical formula: $KAlSi_3O_8$ (potassium, aluminum silicate).

Pyrite—Pyrite is a brassy-yellow mineral with a greenish to brownish-black streak, and has a metallic luster, a specific gravity of 5.02 (high), and a hardness of 6–6.5. Pyrite has cubic or octahedral crystals and striations may be seen on some crystal faces. Chemical formula: FeS_2 (iron sulfide).

Quartz—Quartz is colorless to white but is often tinted. Quartz has a vitreous luster, conchoidal fracture, a specific gravity of 2.65, and a hardness of 7 on Mohs' scale. Chemical formula: SiO_2 (silicon dioxide). Quartz has many varieties.

> **Amethyst**—Amethyst is the purple-tinted hexagonal crystal variety of quartz. See quartz (above) for other properties and chemical formula.

> **Chalcedony/Agate**—Chalcedony is a milky-colored cryptocrystalline variety of quartz. Chalcedony is frequently banded, and more transparent varieties with darker mineral inclusions ("growths") are usually called agate. Chalcedony/agate has a waxy to vitreous luster, and an obvious conchoidal fracture. See quartz (above) for other properties and chemical formula.

Chert/Flint—Chert/flint is an opaque, cryptocrystalline, and darker variety of quartz. Chert is generally lighter in color than flint. The dark-gray to black variety is usually called flint. Chert/flint has waxy to vitreous luster, and an obvious conchoidal fracture. See quartz (above) for other properties and chemical formula.

Milky quartz—Milky quartz is the translucent to white, crystalline variety of quartz with microscopic conchoidal fracture. See quartz (above) for other properties and chemical formula.

Rock crystal—Clear quartz crystals are bipyramidal hexagonal, and usually show striations. See quartz (above) for other properties and chemical formula.

Rose quartz—Rose quartz is the pink-tinted crystalline variety of quartz. See quartz (above) for other properties and chemical formula.

Smoky quartz – Smokey quartz is the smoky-yellow, to brown, to gray, to black variety of crystalline quartz. See quartz (above) for other properties and chemical formula.

Sphalerite—Sphalerite is brown, yellow, or black, has a brown to yellow streak (strong sulfur smell), a resinous to submetallic luster, a specific gravity of 4, and a hardness of 3.5–4. Sphalerite has a perfect dodecahedral cleavage. Chemical formula: ZnS (zinc sulfide).

Sulfur—Sulfur is usually bright yellow but may vary with impurities to green, gray, or red. Sulfur has a white to pale-yellow streak, a resinous to greasy luster, no cleavage, a conchoidal to uneven fracture, a specific gravity of 2, and a hardness of 1.5–2.5. Sulfur has a "rotten egg" odor. Chemical formula: S (sulfur).

Talc—Talc is white, brownish, gray, or greenish-white, and has a white streak, a pearly to dull luster, a specific gravity of 2.7–2.8, and a hardness of 1 on Mohs' scale. Talc has perfect basal cleavage (not apparent in massive specimens), and a smooth or soapy feel. Chemical formula: $Mg_3Si_4O_{10}(OH)_2$ (hydrous magnesium silicate).

*Topaz—Topaz is colorless to white to yellow-brown, and has a white streak, a vitreous luster, a specific gravity of 3.5–3.6, and a hardness of 8 on Mohs' scale. Topaz has perfect basal cleavage (not apparent in massive specimens), and is usually found in coarse crystal masses. Topaz crystals may exhibit striations. Chemical formula: $Al_2SiO_4(OH,F)_2$ (hydrous aluminum silicate).

MINERAL USES LIST

Apatite—Apatite is a phosphate used for fertilizer. Colored varieties are occasionally used as gemstones. Name derivation: Greek *apate*, meaning "deceit," because it was easily confused with other minerals.

Augite—Most augite is only of interest to mineral collectors. Clear varieties are occasionally used as gemstones. Name derivation: Greek *augites*, meaning "brightness" or "luster."

Biotite—Biotite has no economic use but is of interest to collectors. Name derivation: for French physicist J. B. Biot.

Calcite—Calcite has many uses: lime (Ca oxide) is a fertilizer, the raw material from which Portland cement (for making concrete) is made, and a building stone (limestone and marble). Name derivation: Latin *calx*, meaning "burnt lime."

Chlorite—Chlorite has no commercial value, but is a natural green pigment often found in marbles. Name derivation: from Greek *chloros*, meaning "green."

Corundum—Because of its great hardness (9), corundum is used as an abrasive ("black" sandpaper), and for emery wheels for the grinding of metal. Rubies (if red)

* Minerals specimens may not be presented in the laboratory.

and sapphires (if blue, pink, or yellow) are transparent varieties. Name derivation: *kauruntaka*, Indian (Hindu) name for corundum.

Dolomite—Because dolomite contains magnesium, it is a source of this element for magnesium-deficient diets. It is also used as a building stone or as road gravel. Name derivation: after French scientist D. de Dolomieu.

Fluorite—Fluorite is a source of fluorine, used to fluoridate drinking water or added to toothpaste to increase the hardness of dental enamel; it is used in the manufacture of hydrofluoric acid (the only acid that will dissolve glass), and as a flux in steel making and other processes. Name derivation: Latin *fluere*, meaning "to flow," which refers to the ease with which fluorite melts when heated, compared to other minerals.

Galena—Galena is a source of lead as metal when refined, and is used in glass making (leaded crystal) and radiation-shielding material. Name derivation: Latin *galene*, original name for lead ore.

Garnet—Garnet is slightly harder than quartz, and thus is a good abrasive ("red" sandpaper). It is used as a sandblasting medium and as a grit and powder for optical grinding and polishing. When transparent and without internal fractures, garnet is also a semi-precious gem. Name derivation: Latin *granatus*, meaning "like a grain."

Goethite—Goethite is an ore of iron. Name derivation: after J. W. Goethe, a German poet and scientist.

Graphite—Graphite is the "lead" in pencils, a dry lubricant, and is used in the steel industry. Name derivation: Greek *graphein*, meaning "to write."

Gypsum—When the H_2O is driven off by heat, gypsum becomes anhydrite, and when ground to a powder, it becomes plaster of Paris. Gypsum is used in the manufacture of sheet rock, plaster, and plaster casts, and as a fertilizer, among other uses. The alabaster variety is used to make statuary, and satin spar is used as ornamental decoration. Name derivation: Arabic *jibs*, meaning "plaster."

Halite—Used as table salt, a food preservative, for tanning leather, and as a source of sodium and chlorine, among other uses. Name derivation: Greek *halos*, meaning "salt."

Hematite—Hematite is an ore of iron, the material from which, through the smelting process, iron is extracted as pure metal. Hematite ores can run up to about 70% (by weight) iron. Hematite is also used as a red pigment in paint. Name derivation: Greek *haimatos*, meaning "blood," for the blood-red streak color.

Hornblende—Hornblende has no commercial value, but is of interest to collectors. Name derivation: from German *horn* and *blenden*, meaning "horn" and "blind," respectively, in reference to its luster and lack of value.

Kaolinite—Kaolinite is pure china clay, clay for ceramics, and is used as a filler in paper, rubber, candy, medicines, and so forth. Name derivation: Chinese name *Kao-ling*, meaning "high ridge," referring to the area in China where it was first obtained for export.

Magnetite—Magnetite is the most superior iron ore because of its high iron content. Name derivation: from Magnesia, an area near Macedonia, near Greece, where it was originally found.

Malachite—Malachite is an ore of copper and is also used as an ornamental collectible stone. Name derivation: Greek.

Muscovite—Because muscovite is a transparent heat-resistant mineral, it is used as the "windows" in high-temperature ovens. It is also used as an electrical insulator, and was earlier used as decorative "snow" for Christmas ornaments. Name derivation: from the Muscovy area in Russia where it was used as window glass and from Latin *micare*, meaning "to shine."

Olivine—Olivine, because it is heat resistant, is used as a "brick" liner for high-temperature ovens and furnaces. When transparent, it is the gem peridot. Name derivation: from its olive-green color.

Orthoclase—When ground to a powder and mixed with water, orthoclase forms the coating on ceramics that, when fired in a kiln, turns to glaze, or glass. Name derivation: Greek *orthos*, meaning "right angle," and *klasis*, meaning "to break."

Plagioclase Feldspar—Labradorite is used as an ornamental stone when it displays labradoresence (play of colors). Albite, when opalescent, is cut and polished and known as the gem *moonstone*. Name derivation: Greek *plagios*, meaning oblique (cleavage angle).

Pyrite—Because of its high sulfur content, it is used in the manufacture of sulfuric acid. Name derivation: Greek word *pyr*, meaning "fire."

Quartz—Varities include: citrine (yellow), rose (pink), amethyst (purple), smoky (brown-black), milky (white), chalcedony—agate (banded), jasper (red), chert (light gray), flint (dark, dull color), rock crystal (crystal form), and others. Quartz crystals are often used as semi-precious gems, or for display in mineral collections. Agate, if partially transparent or translucent, is often polished and used as a semi-precious gem. Chert and flint are the raw materials from which stone tools were once made. Pure quartz sand is used to make glass. Name derivation: German *quartz*.

Sphalerite—Sphalerite is zinc ore, the material that, when refined, gives us zinc as metal. A thin coating of zinc on iron or steel offers considerable protection from oxidation (rusting). Originally, the zinc was applied by electrolysis, which gave rise to the name "galvanized iron," but it is cheaper to dip the material in a bath of molten zinc. Zinc is used to galvanize corrugated iron roofing, iron buckets, nails, and other materials. Name derivation: Greek *sphaleros*, meaning "treacherous."

Sulfur—Sulfur has many uses. It is used in the manufacture of sulfuric acid. Also, when added to rubber (vulcanized rubber), it makes the rubber able to withstand high temperatures and thus suitable for tires, rubber hoses, and so forth. It is also used in the production of sulfa drugs. Name derivation: from *sulphur*, meaning "brimstone."

Talc—Talc, when ground to a powder and scented, is used as body powder (talcum, baby powder), and as an ingredient in paint, paper, and other materials. Name derivation: Arabic word *talq*, meaning "pure."

Topaz—Topaz is used as a gemstone.

EXERCISE 1: IDENTIFICATION OF MINERAL UNKNOWNS AND THEIR PROPERTIES

LABORATORY MATERIALS

Glass plate	Paper towels	Steel nail	Penny
Streak plate	Magnet	Dilute hydrochloric acid	

The identification of mineral unknowns is determined, by the beginning geology student, by filling in the physical properties of each mineral on the mineral data sheets. Use the tools provided to test the physical properties of each mineral. After determining the physical properties, go to the **Guide for the Identification of White or Light-Colored Minerals** and/or the **Guide for the Identification of Dark or Metallic Minerals**. Match the distinctive set of physical properties of the mineral unknowns to the properties on the guides to identify each mineral. The student may also utilize **Figure 1.3,** the **Mineral Identification Key.**

Do NOT leave any blanks on the mineral data sheets. Blanks will be counted as wrong answers.

GENERAL SAFETY INFORMATION

The identification of mineral unknowns requires the use of materials that can cause injury if used improperly. The following instructions are intended to familiarize the student with the proper laboratory procedures and safety practices.

Glass Plates and Streak Plates

The purpose of the glass plate is to determine the hardness of a mineral unknown. Minerals greater than 5.5 on Mohs' scale of hardness will scratch a glass plate. Hold the glass plate **flat on the table** and scratch the mineral firmly against it. Avoid the edges of the plate. **DO NOT HOLD THE GLASS or STREAK PLATE IN YOUR HAND WHILE SCRATCHING IT**. It can break and cause injury.

Steel Nails

Nails have a sharp, pointy end. When not in use, the nail should be lying flat on the table or in its proper container. If an injury occurs, clean it, bandage it, and see your doctor for a **tetanus shot**, if necessary.

Fingernail

DO NOT scratch your fingernail with the mineral.

Dilute Hydrochloric Acid

Protective goggles should be worn at all times when using HCl acid. Apply ONE drop to the sample, note the reaction (if any), and wipe the acid off the mineral with a paper towel. DO NOT wipe your eyes or put your fingers in your nasal or oral cavities before washing your hands at the end of lab class.

The instructor will point out the Eye Wash Station in the classroom.

FOOD and DRINKS are NOT allowed in lab classrooms.

Properties / Mineral Name					
Chemical Formula					
Hardness (Exact if on Mohs' scale)					
Luster √ = Metallic Describe if nonmetallic					
Streak (color)					
Diaphaneity Transparent, translucent, or opaque					
Magnetism √ = Magnetic X = Nonmagnetic					
Crystal Form/Habit Describe if present X = none					
Cleavage Number of planes X = none					
Specific Gravity H = Heavy, N = Normal L = Lightweight					
Fracture Yes, No, or C = Conchoidal					
Reaction to HCl Describe reaction X if none					
Striations √ = Present X = None					
Color of Mineral					
List one use					

Properties					
Chemical Formula					
Hardness (Exact if on Mohs' scale)					
Luster √ = Metallic Describe if nonmetallic					
Streak (color)					
Diaphaneity Transparent, translucent, or opaque					
Magnetism √ = Magnetic X = Nonmagnetic					
Crystal Form/Habit Describe if present X = none					
Cleavage Number of planes X = none					
Specific Gravity H = Heavy, N = Normal L = Lightweight					
Fracture Yes, No, or C = Conchoidal					
Reaction to HCl Describe reaction X if none					
Striations √ = Present X = None					
Color of Mineral					
List one use					
Other identifiable features					

Properties \ Mineral Name						
Chemical Formula						
Hardness (Exact if on Mohs' scale)						
Luster √ = Metallic Describe if nonmetallic						
Streak (color)						
Diaphaneity Transparent, translucent, or opaque						
Magnetism √ = Magnetic X = Nonmagnetic						
Crystal Form/Habit Describe if present X = none						
Cleavage Number of planes X = none						
Specific Gravity H = Heavy, N = Normal L = Lightweight						
Fracture Yes, No, or C = Conchoidal						
Reaction to HCl Describe reaction X if none						
Striations √ = Present X = None						
Color of Mineral						
List one use						

Properties					
Chemical Formula					
Hardness (Exact if on Mohs' scale)					
Luster √ = Metallic Describe if nonmetallic					
Streak (color)					
Diaphaneity Transparent, translucent, or opaque					
Magnetism √ = Magnetic X = Nonmagnetic					
Crystal Form/Habit Describe if present X = none					
Cleavage Number of planes X = none					
Specific Gravity H = Heavy, N = Normal L = Lightweight					
Fracture Yes, No, or C = Conchoidal					
Reaction to HCl Describe reaction X if none					
Striations √ = Present X = None					
Color of Mineral					
List one use					
Other identifiable features					

Properties \ Mineral Name					
Chemical Formula					
Hardness (Exact if on Mohs' scale)					
Luster √ = Metallic Describe if nonmetallic					
Streak (color)					
Diaphaneity Transparent, translucent, or opaque					
Magnetism √ = Magnetic X = Nonmagnetic					
Crystal Form/Habit Describe if present X = none					
Cleavage Number of planes X = none					
Specific Gravity H = Heavy, N = Normal L = Lightweight					
Fracture Yes, No, or C = Conchoidal					
Reaction to HCl Describe reaction X if none					
Striations √ = Present X = None					
Color of Mineral					
List one use					

Properties / Name					
Chemical Formula					
Hardness (Exact if on Mohs' scale)					
Luster √ = Metallic Describe if nonmetallic					
Streak (color)					
Diaphaneity Transparent, translucent, or opaque					
Magnetism √ = Magnetic X = Nonmagnetic					
Crystal Form/Habit Describe if present X = none					
Cleavage Number of planes X = none					
Specific Gravity H = Heavy, N = Normal L = Lightweight					
Fracture Yes, No, or C = Conchoidal					
Reaction to HCl Describe reaction X if none					
Striations √ = Present X = None					
Color of Mineral					
List one use					
Other identifiable features					

Properties \ Mineral Name					
Chemical Formula					
Hardness (Exact if on Mohs' scale)					
Luster √ = Metallic Describe if nonmetallic					
Streak (color)					
Diaphaneity Transparent, translucent, or opaque					
Magnetism √ = Magnetic X = Nonmagnetic					
Crystal Form/Habit Describe if present X = none					
Cleavage Number of planes X = none					
Specific Gravity H = Heavy, N = Normal L = Lightweight					
Fracture Yes, No, or C = Conchoidal					
Reaction to HCl Describe reaction X if none					
Striations √ = Present X = None					
Color of Mineral					
List one use					

Properties	Name				
Chemical Formula					
Hardness (Exact if on Mohs' scale)					
Luster √ = Metallic Describe if nonmetallic					
Streak (color)					
Diaphaneity Transparent, translucent, or opaque					
Magnetism √ = Magnetic X = Nonmagnetic					
Crystal Form/Habit Describe if present X = none					
Cleavage Number of planes X = none					
Specific Gravity H = Heavy, N = Normal L = Lightweight					
Fracture Yes, No, or C = Conchoidal					
Reaction to HCl Describe reaction X if none					
Striations √ = Present X = None					
Color of Mineral					
List one use					
Other identifiable features					

IGNEOUS ROCK PRELAB WORKSHEET

Name: _____   **Sect.:** _____

Answer the following, as completely as possible.

Define igneous rock. _____

_____

_____

Igneous rocks are classified on the basis of _____ and _____

Define felsic. _____

_____

_____

Provide six examples of felsic rocks.

1. _____   2. _____

3. _____   4. _____

5. _____   6. _____

Define intermediate. _____

_____

_____

Provide two examples of intermediate rocks. 1. _____ 2. _____

Define mafic. _____

_____

_____

Provide two examples of mafic rocks. 1. _____ 2. _____

Define ultramafic. _____

_____

_____

Provide two examples of ultramafic rocks. 1. _____ 2. _____

List and define the six igneous rock textures.

1. _____ - _____

_____

_____

2. _____ - _____

_____

_____

3. _____ - _____

_____

_____

4. _____ - _____

_____

_____

5. _____ - _____

_____

_____

6. _____ - _____

_____

_____

Define magma. _____

_____

Define intrusive (plutonic). _____

_____

Define lava. _____

_____

Define extrusive (volcanic). _____

_____

Define sill. _____

_____

Define dike. _____

_____

Define batholith. _____

_____

Define laccolith. _____

_____

Igneous Rocks

ROCKS

A **rock** is a natural aggregate (combination) of one or more minerals, mineraloids, and/or organic material. There are three families of rocks distinguished from one another by the processes involved in their formation:

1. **Igneous rocks** originate from a molten material (lava or magma).
2. **Sedimentary rocks** originate by the deposition of the by-products of weathering.
3. **Metamorphic rocks** originate via the change of preexisting rock by heat, pressure, and/or the addition of hot chemical fluids.

Igneous, sedimentary, and metamorphic rocks are described and identified on the basis of composition and texture. **Composition**, in general, refers to the chemical makeup, the particular elements in the rock. **Texture**, in general, refers to the size, shape, and arrangement of the constituent minerals or materials that make up the rock. There are different sets of textural terms for each rock family that often denote the same or closely similar conditions.

IGNEOUS ROCKS

Igneous rocks are the solids produced by the cooling and crystallization of molten silicate material initially formed beneath the Earth's surface. Crystallization occurs when cooling allows for the growth of mineral crystal grains. The cooling rate and space available determine the size of the crystals that form. Large crystals form when **magma**, molten silicate material below ground, is insulated by the surrounding country rock (preexisting rock that has been intruded by the magma), and therefore cools very slowly. Igneous rocks that form when magma cools underground are referred to as **plutonic (intrusive)** igneous rocks. The shape and position of emplacement differentiate plutonic igneous rock bodies. A relatively small two-dimensional pluton that cuts across preexisting country rock is a **dike**. *(Principle of Cross-Cutting Relationships: A rock body must already exist in order for it to be cut by another)*. A **sill** is a two-dimensional pluton that is emplaced parallel to and between layers (strata) of preexisting rock. A **laccolith** is a three-dimensional pluton with a convex roof and a flat floor. **Batholiths** are very large, three-dimensional plutons, usually the result of multiple intrusions of magma, hundreds of miles in length and width, that cool and crystallize very slowly beneath the Earth's surface.

Volcanic (extrusive) igneous rocks form on or above the surface of the Earth by the cooling of **lava** (molten silicate material on the surface), or by the deposition of violently ejected **pyroclastic** (*pyro* = fire, *clast* = fragment) material such as volcanic ash. Lava cools much faster than magma because it is exposed to environments that allow for rapid dissipation of heat, therefore preventing the formation of large crystals. In general, most extrusive igneous rocks develop crystals that are too small to be seen without the aid of a microscope. There are different types of basaltic lava flows. **Aa** is a blocky, sharp-edged lava that moves very slowly, and **pahoehoe** is a ropy, "smooth" lava. **Obsidian** (volcanic glass) forms when lava is cooled too rapidly for crystals to develop.

Bowen's Reaction Series

Igneous rocks, with few exceptions, are composed of silicate minerals. An understanding of igneous rock formation can be gained by considering Bowen's Reaction Series. Bowen's Reaction Series (Figure 2.1) is the result of experiments conducted by N. L. Bowen and first published in 1928. **Bowen's Reaction Series** is an organization of the silicate minerals according to the conditions required to crystallize them, as the temperature of a melt lowers. Bowen discovered that in addition to the availability of needed chemical elements, temperature and pressure determine when and where given silicate minerals form. He observed that some minerals form as a **continuous series** belonging to a single silicate family (tectosilicates) but with a progressive change (substitution) of chemical composition, whereas others form as a **discontinuous series** of different silicate crystal families as their crystal structures readjust. The discontinuous series of readjustments proceeds from what can be thought of as zero-dimensional arrangements (highest temperatures and pressures) through 1-D, 2-D, to 3-D arrangements (low temperature and pressure), *if all of the necessary elements to build a particular mineral are available.*

The continuous series involves the plagioclase feldspar group. These minerals have a three-dimensional covalently bonded structure that includes metal ions. The structure is

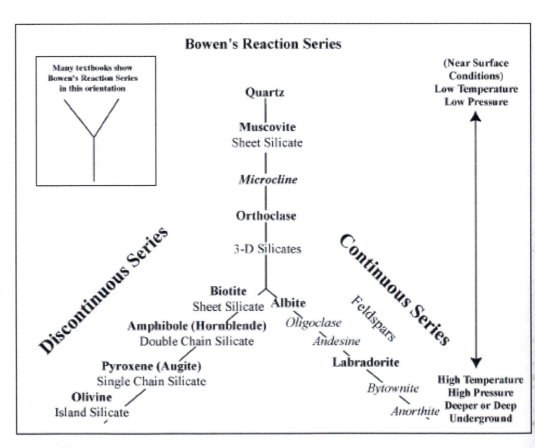

FIGURE 2.1 Bowen's Reaction Series.

continuously modified as ions are exchanged with the magma during cooling. Calcium-rich (Ca) plagioclase crystals (anorthite, $CaAl_2Si_2O_8$) first begin to form when the magma has cooled to 1400° to 1200°C. As cooling continues (1200° to 1000°C), the crystals exchange calcium and aluminum (Al) ions for sodium (Na) and silicon (Si) ions from the magma, to form crystals that are more sodium- and silicon-rich. Calcium-rich plagioclase crystals also form directly from the magma at this temperature range. If the temperature of the magma continues to decrease very slowly, so that equilibrium is approximately maintained, plagioclase feldspars will continue to exchange ions in this manner until the magma solidifies. If there is sufficient sodium, Ca-plagioclases disappear completely, but in many magmas, all of the Na and Al becomes bonded early and is lost from the system. Thus this process—which can proceed successively from anorthite (Ca-rich), to bytownite, labradorite, andesine, oligoclase, and albite (Na-rich)—in practice produces a variety of different minerals, depending on the original composition of the magma and the rate of cooling.

Silicate minerals of the discontinuous series have a variety of different structures of increasing complication that appear and disappear successively and predictably, as conditions (mainly temperature) in magmas change. The following discussion is primarily concerned with decreasing temperature, but the effects of pressure are generally similar. Olivine is the first mineral (stable silicate or structure) to appear at 1400° to 1200°C. The olivine crystal consists of individual tetrahedra (plural of *tetrahedron*; four oxygen and a much smaller silicon hidden in the center; Figure 2.2a) tied together by bivalent iron (Fe^{++}) and magnesium (Mg^{++}) ions in a three-dimensional network. Olivine crystals become unstable when the melt cools to about 1200° to 1000°C, the temperature range in which pyroxene becomes stable. Augite is an example of a common mineral in the pyroxene family. Olivine crystals suspended in the magma react to form a more complex single chain (pyroxene, augite; Figure 2.2b) silicate structure. Amphibole (hornblende) becomes stable at approximately 1000° to 800°C. Again, the earlier-formed pyroxene (augite) crystals react with the melt and form double-chain (Figure 2.2c) amphibole (hornblende) crystal structures. If sufficient magma and silica (SiO_2) are still available, the hornblende will react with it and change to biotite, a sheet silicate (Figure 2.2d). Orthoclase and microcline (both three-dimensional covalently bonded structures with metal ions), muscovite (sheet silicate), and quartz (three-dimensional structure) will form last if enough magma is left.

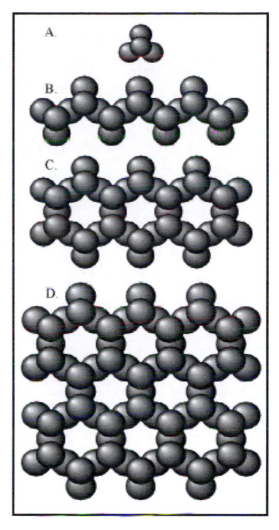

FIGURE 2.2 Silicate Structures.

Igneous Rocks: Composition

The composition of igneous rocks can be determined, in a general way, in hand specimens by the relative abundance and color intensity (pale versus dark or strong colored) of the minerals that make up the rock. Although the composition of many igneous rocks can be determined in this way, many specimens are not so easily classified. Knowledge of the

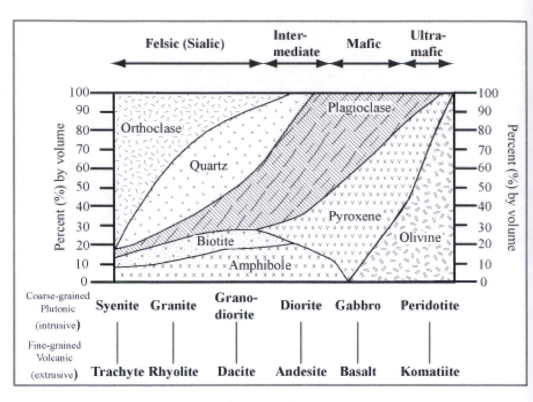

FIGURE 2.3 Common Igneous Rocks (mineralogy and composition).

mineral constituents and the percentage of each mineral within the rock is also used to determine the composition. Igneous rocks are divided into four compositional ranges, as illustrated in Figure 2.3:

1. **Felsic composition**—Igneous rocks composed mainly of potassic and sodic feldspars (light-colored minerals) that formed under low-temperature and low-pressure conditions (Bowen's Reaction Series). Felsic rocks include syenite and trachyte, *granite* and *rhyolite*, granodiorite and dacite, and some *obsidian*.

2. **Intermediate composition**—Igneous rocks with subequal amounts of light- and dark-colored minerals. *Diorite* and *andesite* (named for the Andes Mountains), and some *obsidian*, are intermediate in composition.

3. **Mafic composition**—Igneous rocks that have a large percentage of darker and strongly colored minerals rich in ferromagnesian components and calcic plagioclase feldspars. These minerals form under high-temperature and high-pressure conditions (Bowen's Reaction Series). *Gabbro* and *basalt*, and some *obsidian*, are mafic in composition.

4. **Ultramafic composition**—Igneous rocks that often contain 70 to 90% olivine, other dark and strongly colored ferromagnesian minerals, the most calcic plagioclases, and very minor, if any, silica. These minerals form under very high-temperature and high-pressure conditions (Bowen's Reaction Series). Ultramafic rocks include *peridotite* and *komatiite*. Ultramafic rocks are not common at or near the Earth's surface, but form in the asthenosphere and mantle.

Igneous Rock Texture

The **texture** of igneous rocks refers to the physical appearance ("visual feel") and arrangement of minerals within the rock. Texture may include the absence of crystals in a rock, the presence and/or relative size of the crystals that make up the rock, any contrast in crystal sizes within a given rock specimen, the arrangement of minerals in the rock, and/or the presence of bubbles (**vesicles**) in the rock.

1. **Glassy texture** is applied to igneous rocks that have cooled so rapidly that crystals didn't have time to develop or grow. **Vitrophyres** are igneous rocks that are a combination of glass

and visible crystals. Vitrophyres often form during rapid intrusion when magma comes in contact with the much cooler surrounding country rock. A **chilled** (rapidly cooled) **margin** is a thin zone of rapidly cooled igneous rock that forms a rind on the pluton, and can be aphanitic, glassy, or vitrophyric. *Obsidian*, volcanic glass, has a glassy texture.

2. **Aphanitic** is a textural term used to describe igneous rocks that have crystals that are approximately uniform and small in size. "Small" means that the crystals are not distinguishable by the unaided eye (<<1 millimeter). Both **microcrystalline** (crystals visible only under a microscope) and **cryptocrystalline** (crystals too small to be seen with the ordinary microscope) textures are included.

3. **Phaneritic** is the textural term used to describe igneous rocks that have crystals large enough to be seen without magnification, 1 mm to 2.54 cm ($\approx$1 inch), or medium size.

4. **Pegmatitic** is the textural term applied to igneous rocks in which the crystals are large or very large. "Large" means *very coarsely crystalline*. The crystals in a pegmatite may be a few centimeters (1 inch) to several meters (100 cm, or $\approx$39.33 in/m) in length.

5. **Porphyritic** is the textural term used when the crystals in the rock fall into two distinct size groups (small versus large). The larger (phaneritic) crystals are surrounded by a finer-grained matrix. Rocks with this combination of crystal sizes usually have the word *porphyritic* appended to the rock name. *Porphyritic basalt* is an example.

   A **vitrophyre** is a special kind of porphyritic rock. The smaller aphanitic crystals and glass, if present, form what is called the **matrix** or **groundmass matrix** of the rock. The matrix, if aphanitic or partly glassy, contains minerals formed at low temperatures and pressures. The larger crystals are called **phenocrysts**. Phenocrysts are commonly early-formed, slow-growing minerals that crystallize at higher temperatures and pressures. As a result, phenocrysts are more likely to have better-formed crystal faces (be euhedral) than crystals that form later. Crystals that form late, quartz in a granite for instance, do not develop crystal faces (are anhedral). The rock is probably volcanic or part of a very shallow intrusion (plumbing of a volcano) if the groundmass is aphanitic.

6. **Vesicular** texture describes a volcanic igneous rock with bubbles (holes). The bubbles form when pressure is released during eruption and volatile components of a magma exsolve (come out of solution). Water (H_2O) and carbon dioxide (CO_2) are the two most abundant volatile components.

Bubbles are most commonly found in volcanic rocks, but sometimes occur in the uppermost parts of dikes that were part of the plumbing for an eruptive center. Expansion of the gases formed the bubbles, and the expansion helps to cool the magma/lava. Bubbles range in size from very small (small fractions of a millimeter) to more than a meter, although very large vesicles are uncommon. The rate of cooling and the viscosity of the magma/lava control the size of the bubbles. Most of the terms arising from vesicular textures are associated with abundance of vesicles as well as size. Rocks with *widely spaced and clearly visible vesicles* are referred to as vesicular. *Vesicular basalt* is a common volcanic igneous rock.

Rocks with *closely spaced bubbles* that are on the order of 1 to 2 mm in diameter or larger are referred to as **scoria**. Scoriaceous basalt identifies the tops of basalt flows. Contact with oxygen in the air oxidizes iron in the glass to produce tiny crystals of hematite, and a reddish color in many scorias. Scoria has a very low density for a rock, but usually does not float on water. *Most scoriaceous rocks are mafic in composition.*

Rocks with closely spaced, microscopic bubbles (less than a millimeter, usually) may be referred to as **pumice**. *Most pumiceous rocks are felsic in composition.* Pumice often has a density so low that the rock will float on water. Both scoria and pumice have vesicular texture, yet both are forms of obsidian, volcanic glass. The release of gases creates a frothy or vesicular texture in the obsidian.

Vesicles later filled with a solid material (secondary minerals) are called **amygdules**. *Amygdaloidal basalt* is basalt with filled vesicles. Common secondary minerals include quartz crystals, chalcedony, agate, and calcite. Large chalcedony or agate-filled amygdules can be handsomely colored and have some value to collectors. Large, partially hollow amygdules are sometimes referred to as "geodes," though technically geodes form in sedimentary rocks.

ARRANGEMENTS OF CRYSTALS AND BUBBLES

The **arrangement of crystals and bubbles** is also an important aspect of texture in igneous rocks. For the purpose of this discussion, an "arrangement" of textural elements is a situation where the occurrence or orientation of the feature is not random. Arrangements are a product of local variations in the chemicals and physical conditions within the magma or lava, and gravity. Understanding of such nonrandom organizations of textural features, and particularly specific kinds of crystals (minerals), is a major area of study in *igneous petrology* and *geochemistry*.

The beginning of crystallization of any mineral variety requires just the right balance of ion availability, temperature, and pressure. A simple way for all of these to vary at once is for the magma to be in motion, flowing. Shearing stresses induced by flow can align existing **acicular** (needle-shaped) crystals, producing an arrangement called **flow lineation**. Surfaces of low pressure sub-parallel to the boundaries of flow develop when flow expands (cross-sectional area of flow increases). These can localize precipitation of sheet-like mineral masses, **flow foliation**. Early-formed crystals have densities greater than that of the magma, and collect near the bottom of magma chambers under the influence of gravity.

The natural form of bubbles and vesicles is **spherical**, or, when many bubbles are closely packed (scoria), compact. Under the influence of gravity, bubbles rise and expand to collect at the top of a lava flow, if the viscosity is low. Flow will deform vesicles into **ellipsoids,** all aligned in the direction of flow. Basaltic lava can move fast enough to achieve very complicated (turbulent) flow, somewhat like that of water in a brook. In this situation, vesicles can take complicated shapes.

How to Use Figure 2.3 Common Igneous Rocks (Mineralogy and Composition)

Figure 2.3 is used to determine the composition, mineral percentage, and an initial crystal texture (coarse—large visible crystals, versus fine—small crystals not distinguishable to the naked eye) of unknown igneous rocks. The rocks in the bottom row (trachyte–komatiite) are the result of volcanism (extrusive igneous rocks) and therefore cooled too fast for large crystals to form. The upper row of rocks (syenite–peridotite) cooled slowly underground (plutonic/intrusive igneous rocks) and therefore large visible crystals developed.

The composition of the rocks is determined by extending the line between any set of rocks (e.g., rhyolite/granite) upward until it intersects the compositional terms at the top of the graph.

The percentage of each mineral in the rocks can also be determined by extending the line between any two rocks upward. First, determine which minerals intersect this line. Granodiorite/dacite, for example, contains amphibole, biotite, plagioclase, quartz, and orthoclase.

Amphibole begins at 0% and changes into biotite at 18%. Biotite begins at 18% and changes to plagioclase at 26%. Plagioclase begins at 26% and changes to quartz at 50%. Quartz begins at 50% and changes to orthoclase at 89%. Orthoclase begins at 89% and ends at 100%. Therefore, both granodiorite and dacite contain the following percentages of minerals:

18% amphibole	18%–0% = 18%
8% biotite	26%–18% = 8%
24% plagioclase	50%–26% = 24%
39% quartz	89%–50% = 39%
11% orthoclase	100%–89% = 11%

The only difference between the granodiorite and dacite is the size of the minerals. Granodiorite has large, visible crystals (phaneritic or porphyritic) and dacite has small, not visible crystals (aphanitic, in general).

Igneous rock pictures and other information can be found on the Earth and Space Science website (http://ess.lamar.edu). Click on the People tab, Staff, Karen M. Woods, Teaching, Physical Geology Lab, Rocks, Igneous Rocks.

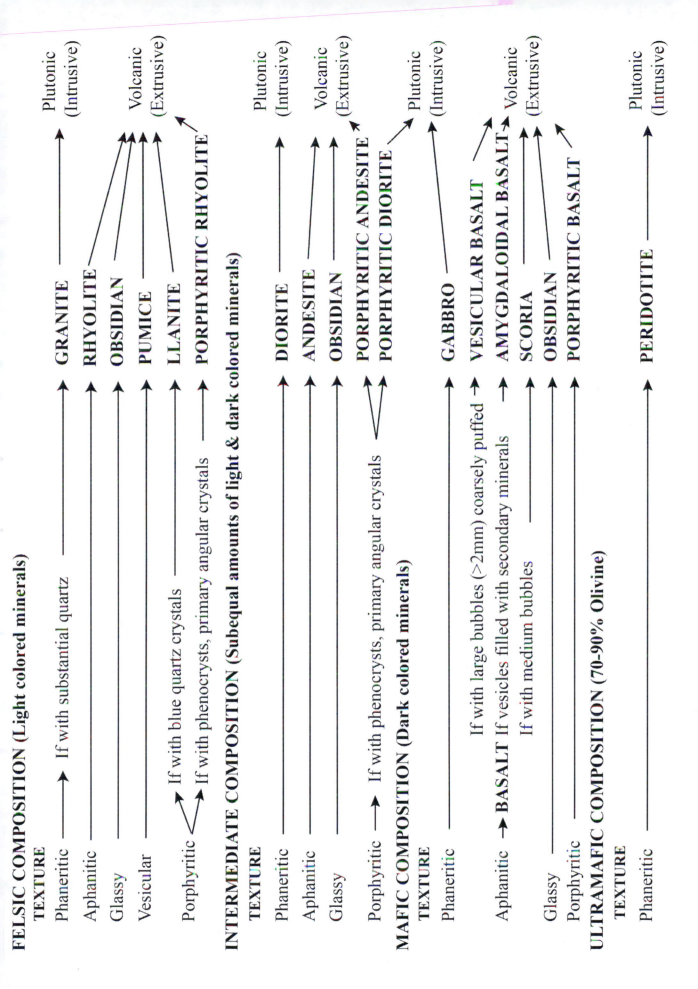

FELSIC COMPOSITION (Light colored minerals)
TEXTURE
Phaneritic → If with substantial quartz → GRANITE → Plutonic (Intrusive)
Aphanitic → RHYOLITE → Volcanic (Extrusive)
Glassy → OBSIDIAN
Vesicular → PUMICE
Porphyritic:
 If with blue quartz crystals → LLANITE
 If with phenocrysts, primary angular crystals → PORPHYRITIC RHYOLITE

INTERMEDIATE COMPOSITION (Subequal amounts of light & dark colored minerals)
TEXTURE
Phaneritic → DIORITE → Plutonic (Intrusive)
Aphanitic → ANDESITE → Volcanic (Extrusive)
Glassy → OBSIDIAN
Porphyritic → If with phenocrysts, primary angular crystals → PORPHYRITIC ANDESITE / PORPHYRITIC DIORITE

MAFIC COMPOSITION (Dark colored minerals)
TEXTURE
Phaneritic → GABBRO → Plutonic (Intrusive)
Aphanitic → BASALT
 If with large bubbles (>2mm) coarsely puffed → VESICULAR BASALT → Volcanic (Extrusive)
 If vesicles filled with secondary minerals → AMYGDALOIDAL BASALT
 If with medium bubbles → SCORIA
Glassy → OBSIDIAN
Porphyritic → PORPHYRITIC BASALT

ULTRAMAFIC COMPOSITION (70-90% Olivine)
TEXTURE
Phaneritic → PERIDOTITE → Plutonic (Intrusive)

37

Guide to the Identification of Common Igneous Rocks

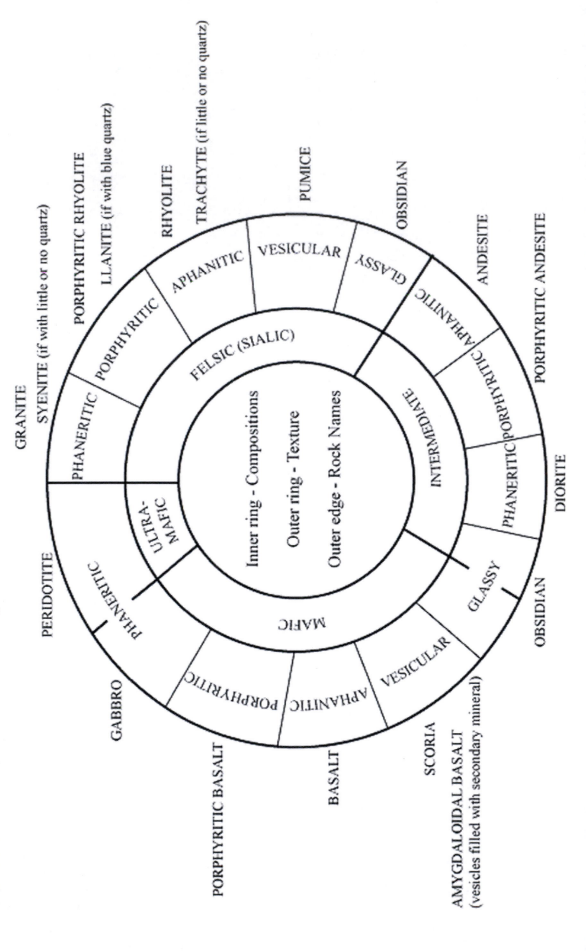

EXERCISE 2.1: IGNEOUS ROCK IDENTIFICATION

Fill in all the blanks in the exercise. DO NOT leave any blanks. If the answer does not apply, write N/A.

The identification of unknown igneous rocks is determined by the visual aspects of each sample.

Composition:

Identify the recognizable minerals within the rock.

Determine the relative amounts of light- and dark-colored minerals within each rock sample.

Use Figure 2.1 to identify the correct compositional terminology for each rock sample.

Texture:

Study each rock sample and determine the general size of the mineral crystals in the rock (very large [coarse], large [coarse], small, both large [coarse] and small, or absent).

Choose the correct term to describe the igneous texture.

Environment of Formation:

Study each sample and determine if the rock cooled for a long time (large crystals) or cooled rapidly (small crystals).

Choose the correct term (intrusive or extrusive) to describe the environment of formation for each rock.

Mineral Percentages:

See explanation of Figure 2.3 previously provided.

Rock Name \ Properties						
Composition						
Texture						
√ if Intrusive X if Extrusive						
% Amphibole						
% Biotite						
% Olivine						
% Orthoclase Feldspar						
% Plagioclase Feldspar						
% Pyroxene						
% Quartz						
Description						

Rock Name / Properties											
Composition											
Texture											
√ if Intrusive X if Extrusive											
% Amphibole											
% Biotite											
% Olivine											
% Orthoclase Feldspar											
% Plagioclase Feldspar											
% Pyroxene											
% Quartz											
Description											

Rock Name / Properties											
Composition											
Texture											
√ if Intrusive X if Extrusive											
% Amphibole											
% Biotite											
% Olivine											
% Orthoclase Feldspar											
% Plagioclase Feldspar											
% Pyroxene											
% Quartz											
Description											

Rock Name / Properties												
Composition												
Texture												
√ if Intrusive X if Extrusive												
% Amphibole												
% Biotite												
% Olivine												
% Orthoclase Feldspar												
% Plagioclase Feldspar												
% Pyroxene												
% Quartz												
Description												

SEDIMENTARY ROCK PRELAB
WORKSHEET

Name: _____   **Sect.:** _____

Define sedimentary rock. _____

_____

_____

_____

What are terrigenous sedimentary rocks? _____

_____

_____

What is a clast? _____

Define clastic texture. _____

_____

How do clasts originate? _____

_____

What is the size range in millimeters of coarse-sized clasts? _____

What is the size range in millimeters of medium-sized clasts? _____

What is the size range in millimeters of fine-sized clasts? _____

What is the size range in millimeters of very-fine-sized clasts? _____

Define conglomerate. _____

_____

Define sandstone. _____

_____

Define siltstone. _____

_____

Define shale. _____

_____

List and define the two ways sediments are lithified.

1. _____ - _____

_____

_____

2. _____ - _____

_____

_____

List the three common agents of cementation.

a. _____ b. _____ c. _____

What are evaporitic sedimentary rocks? _____

_____

_____

Define crystalline texture. _____

_____

_____

Define cryptocrystalline texture. _____

_____

_____

List three common evaporites.

a. _____ b. _____ c. _____

What are biogenic sedimentary rocks? _____

_____

_____

Define bioclastic texture. _____

_____

_____

List the two major kinds of biogenic sedimentary rocks based on their compositions.

a. _____ b. _____

What is the main mineral in rock salt? _____

What is the main mineral in chert? _____

What is coquina composed of? _____

What is fossiliferous limestone composed of? _____

What are peat, lignite, and bituminous coal composed of? _____

Micrite and chert look very similar. What test can be used to distinguish them from one another? _____

SEDIMENTARY ROCKS

Sediments are particles that settle to the bottom of a basin of deposition, such as a lake, river, sea, and so on. **Sedimentary rocks** form as the result of the erosion (picking up), transportation (carrying away), deposition (laying down), and lithification (compaction and/or cementation) of sediments (mechanical processes); some are the result of chemical processes, and some are the result of the accumulation and lithification of organic material (plants or shells). Weathering (rock and mineral destruction) operates continuously at the Earth's surface. The by-products of weathering include the generation of silicate clastic particles (large to small fragments of preexisting rocks), clays, soluble silica, soluble salts (K, Na, Ca, Mg, bicarbonates, etc.), and/or iron oxides.

Unconsolidated sediments are lithified (turned to stone) by two main processes, *compaction* and *cementation*. **Compaction** results when the pore spaces between clasts (grains) are reduced, due to the weight of the overlying sediments (overburden). **Shale** is a very-fine-grained rock with a tendency to part or break in a direction at right angles to the direction of the weight that caused the compaction. If there is no preferred direction of breakage (and thus has conchoidal fracture), the fine-grained, soft rock is called a **claystone**. **Cementation** results when mineral-bearing groundwater moves through the pore spaces of the sediment and deposits dissolved mineral material in the pores. The added mineral, with time, "glues" the grains together. Common cementing agents include **calcite**, **silica**, and **iron oxides**.

Sedimentary rocks are described and named according to their texture, composition, and internal structures directly related to their **mode of formation**. Sedimentary particles originate in three general ways (mode of formation):

1. **Terrigenous**—those derived from the land.
2. **Evaporitic**—crystalline precipitates, those crystallized from an aqueous solution.
3. **Biogenic**—those derived from living matter (plant or animal).

Terrigenous Sedimentary Rock

Terrigenous is the term used to describe material derived from the land. Large to small fragments of preexisting minerals or rocks are called **clasts**. **Clastic** is a textural term used to describe sediment or sedimentary rocks composed of clasts derived from the mechanical or partial chemical weathering of silicate precursors. Pebbles, sand, silt, and clay are terrigenous, extrabasinal (originate outside the basin of deposition) sediments. Lakes, ponds, oceans, and seas are basins of deposition. Intrabasinal sediments originate within the basin of deposition and can be composed of either evaporitic or biogenic material. Because biogenic sedimentary rocks are formed from bits and pieces derived from once-living organisms, they have a bioclastic texture.

The texture of clastic silicate rocks is based on the clast size. There are four categories:

1. **Coarse-grained clastic texture**—the clasts are > 2 mm, includes pebbles, cobbles, and boulders. *Conglomerate* and *breccia* are examples.
2. **Medium-grained clastic texture**—the clasts are 1/16−2 mm (sand-sized). *Sandstone* is an example.
3. **Fine-grained clastic texture**—the clasts are 1/256−1/16 mm (silt-sized). *Siltstone* is an example.
4. **Very-fine-grained clastic texture** —<1/256 mm (clay-sized). *Shale and Claystone* are examples.

The grains that make up silicate clastic sedimentary rocks have an angular, sub-rounded, or rounded shape. Angular-shaped clasts indicate that the clasts either did not travel far from the source area to the basin of deposition or that the clasts are hard and more resistant to erosion. The size, shape, and process of deposition determine the orientation of the grains as they are deposited. Roundness is a partial indicator of the distance sediment is transported by running water before deposition. A variable, however, that influences roundness is the hardness and tenacity of the material being abraded. A coarse sedimentary rock with rounded (even slightly rounded) clasts is called a **conglomerate**, but if the clasts remain angular, the rock is called **breccia**. Breccias can be composed of either igneous or sedimentary rock fragments.

Evaporite Sedimentary Rocks

Evaporites are crystalline precipitates, rocks crystallized from an aqueous solution. They are the second major group of sedimentary rocks and originate in different and often more complex ways. Simple examples of this mode of origin include the chemical precipitation of some dolostone, gypsum, halite, and calcite as mineral-bearing water evaporates on the bottom of the basin of deposition. Most evaporites, such as rock salt and rock gypsum, have a **crystalline texture** (granular) because the atoms of the rock are arranged in an orderly fashion in multiple pieces big enough to see. Rocks such as chert have **cryptocrystalline texture**—the crystals within the rock are too small to be seen without a microscope. The word *cryptocrystalline* means "hidden crystals."

Biogenic Sedimentary Rocks

A third way that sediments (sedimentary rocks) can form is through biologic processes, the by-products of plants or animals. They have a biogenic mode of origin. **Biogenic** sedimentary rocks are made up of clasts derived from once-living organisms. These biologic clasts are referred to as bioclastic; therefore, a **bioclastic** rock is a rock composed of bits and pieces of biogenic material such as shell and plant material. Examples include *peat* (a sediment), *lignite*, and *bituminous coal* (plant accumulations, *coals*), and various kinds of *limestone*. Lignite and bituminous coal are composed of compacted bits and pieces of plant matter and have a coarse-grained bioclastic texture. Most limestone is made up of the accumulation of microscopic or macroscopic skeletons of calcareous marine organisms. When alive, these organisms extracted Ca^+ and $(CO_3)^-$ ions from the water, and used them to form the mineral aragonite ($CaCO_3$) in the construction of their hard parts. When the organisms die, their skeletons (shells) settle to the bottom of the depositional basin and later become compacted or recemented into the sedimentary rocks *micrite, chalk, coquina,* or *fossiliferous limestone*. The same size classification exists for biogenic rocks as with terrigenous rocks (clastic coarse, medium, fine, and very fine). Therefore, the textures become **bioclastic: coarse, bioclastic: medium, bioclastic: fine,** and **bioclastic: very fine.**

SEDIMENTARY STRUCTURES

The deposition of sediments results in features, sedimentary structures, that are useful in the interpretation of the environment the rock was deposited in. **Horizontal layering** is a large-scale primary structure that can be seen in cross sections (such as the walls of a canyon) of any material deposited under the influence of gravity. Within given layers, sedimentary rocks often have thinner layers inclined to the overall horizontal layering called **cross-bedding**. Cross-bedded deposits can be **marine** (deposited in the ocean), **aeolian** (deposited by wind action), or **fluvial** (deposited by stream action) in origin. Small-scale cross-bedding is made by the movement of sediment ripple marks. A **ripple mark** is a ridge-and-trough set that is formed by the action of wind or water. Ripples are either symmetrical or asymmetrical and are developed in aeolian, marine, fluvial, or lacustrine (lake) environments. If a ripple mark is asymmetrical, the flow direction of the paleocurrent can be determined. The less-steep side (stoss) points upstream. The movement of dunes from one to many tens of feet high produces larger-scale cross-bedding that, except for size, often looks similar to that produced by the movement of ripples. A decrease in the grain size of a deposit, from the bottom to the top, is called **graded bedding**.

Composition

The **composition** of sedimentary rocks is based on the chemical composition of the clasts or materials of which they are made (silicate clastic sedimentary rocks), the $CaCO_3$ of limestones (biogenic), the carbon content of coals (biogenic), the minerals within the evaporites ($CaSO_4$, $CaMgCO_3$), and the compositions of the cementing agent.

Sedimentary rock pictures and other information can be found on the Earth and Space Science website (http://ess.lamar.edu). Click on the People tab, Staff, Karen M. Woods, Teaching, Physical Geology Lab, Sedimentary Rocks.

Key to the Identification of Common Sedimentary Rocks

CLASTIC (TERRIGENOUS) SEDIMENTARY ROCKS

Composed of lithified clastic sediments
- Lithified rounded pebbles (>2 mm) → **CONGLOMERATE**
- Lithified angular pebbles (>2 mm) → **BRECCIA**
- Lithified sand grains (1/16-2 mm) → **SANDSTONE**
- Lithified silt (1/16-1/256mm) → **SILTSTONE**
- Lithified mud/clay (<1/256mm) → **SHALE**

BIOGENIC SEDIMENTARY ROCKS

"COALS" Dark brown or black lightweight, burnable
- Visible plant remains, organic debris → **PEAT**
- Compacted, friable, brownish to black → **LIGNITE**
- Hard and black, with layering → **BITUMINOUS COAL**

LIMESTONES Entire rock fizzes easily with application of HCl
- Microscopic fossils, soft, chalky → **CHALK**
- Both microscopic and macroscopic fossils → **FOSSILIFEROUS LIMESTONE**
- Macroscopic fossils, shells and shell fragments → **COQUINA**
- Microscopic fossils, hard with conchoidal fracture → **MICRITE**

EVAPORITIC SEDIMENTARY ROCKS

Chemical precipitate
- Conchoidal Fracture, cryptocrystalline, does NOT react to HCl acid → **CHERT**
- Reacts to HCl acid when powdered → **DOLOSTONE**
- Near white, peach or banded, soft, → **ROCK GYPSUM**
- Salty, glassy (crystalline) texture → **ROCK SALT**

Guide to the Identification of Sedimentary Rocks

Clastic Sediments — Terrigenous Origin

- Boulders — Conglomerate
- Pebbles — Conglomerate
- Coarse sands — Sandstone
- Fine sands — Sandstone
- Silts — Siltstone
- Muds/Clays — Shale

Evaporation of Mineral-Bearing Water — Evaporitic Origin

- Si^{4+} Ions, O^{2-} Ions — Chert (Quartz)
- Ca^+ Ions, HCO_3^- Ions — Cave Formations (Cave Onyx)
- Ca^+ Mg^+ Ions, HCO_3^- Ions — Some Dolostone
- Na^+ Cl^- Ions — Rock Salt (Halite)
- Ca^+ Ions, SO_4^{2-} Ions — Rock Gypsum (Alabaster)
- Satin Spar Gypsum (Fracture fillings)
- Selenite Gypsum (Fracture fillings: Along Bedding Planes)

Organically Generated Material — Biogenic Origin

- Shells or Shell "Hash" — Coquina
- Mixture of (limy) muds and fossils — Fossiferous Limestone
- Calcareous Material
- Calcareous mud — Micrite
- Microfossils — Chalk
- Magnesium bearing muds — Dolostone
- Coral Reef — Limestone
- Plant Debris — Peat-Lignite Bituminous Coal
- Coarse
- Fine

EXERCISE 2.2: SEDIMENTARY ROCK IDENTIFICATION

Fill in all the spaces in the exercise. DO NOT leave any blanks. If the answer does not apply, write N/A.

The identification of unknown sedimentary rocks is determined by the visual and textural aspects of each sample.

Texture:

Study each rock sample and determine the general size of the sediments in the rock and choose the correct term to describe the sedimentary texture.

Composition:

Identify the recognizable sediments within the rock.

Use the Guide to the Identification of Sedimentary Rock and the Key to the identification of Sedimentary Rocks as aids to identify the sedimentary unknowns.

Environment of Formation:

Study each sample and determine if the rock is composed of clastic sediments (terrigenous origin), biologic sediments (biogenic origin), or evaporite sediments (evaporitic origin).

Properties Rock Name	Texture Clastic (specify) Bioclastic (specify) Other (specify)	Composition	Mode of Origin T - Terrigenous B - Biogenic E - Evaporitic	Description

55

Properties / Rock Name	Texture Clastic (specify) Bioclastic (specify) Other (specify)	Composition	Mode of Origin T - Terrigenous B - Biogenic E - Evaporitic	Description

56

METAMORPHIC ROCK PRELAB WORKSHEET

Name: _____   **Sect.:** _____

What does "meta" mean? _____ What does "morph" mean? _____

Define metamorphic rock. _____

_____

_____

Rocks are changed by metamorphism in three ways. What are they?

1. _____ & _____

2. _____ 3. _____

List and define the two classes of metamorphism.

1. _____ - _____

_____

_____

2. _____ - _____

_____

_____

Define contact aureole (halo). _____

_____

List the four metamorphic grades, associated temperature in °C (when listed), pressure, and associated metamorphic facies.

1. Grade _____   Temperature _____

   Pressure _____   Facies _____

2. Grade _____   Temperature _____

   Pressure _____   Facies _____

3. Grade _____   Temperature _____

   Pressure _____   Facies _____

4. Grade _____ Temperature _____

   Pressure _____ Facies _____

Define precursor. _____

_____

What is the igneous precursor of granite gneiss? _____

What is the sedimentary precursor of anthracite coal? _____

What is the metamorphic precursor of graphite schist? _____

What is the sedimentary precursor of skarn? _____ or _____

What is the *immediate* metamorphic precursor of muscovite schist? _____

What is the *igneous* precursor of hornfels? _____

Metamorphic rocks have two major types of texture. List and define both.

1. _____ - _____

_____

_____

2. _____ - _____

_____

_____

List the foliated subtextures and the name of one example (rock name) of each.

Foliated subtexture	Rock name (specify)
1. _____	_____
2. _____	_____
3. _____	_____
4. _____	_____

List the nonfoliated subtextures and the name of one example (rock name) of each.

Nonfoliated subtexture	Rock name
1. _____	_____
2. _____	_____
3. _____	_____

METAMORPHIC ROCKS

Metamorphic rocks are formed by the reaction of preexisting rocks to new conditions of heat, pressure, and/or the presence of hot chemical fluids that move through the rocks. The word *metamorphic* is derived from the Greek words "meta," which means *change*, and "morph," which means *form*. Any preexisting rocks (igneous, sedimentary, or older metamorphic) can be metamorphosed. Metamorphic rocks are classified on the basis of their composition and texture.

Composition

The **composition** of metamorphic rocks is dependent upon the composition of the precursor rock, the conditions of metamorphism, and the content of any migrating fluids. The **precursor rock** is the original unchanged rock or a lower metamorphosed rock and is also referred to as the **parent rock** or **protolith**.

Metamorphism changes precursors in three ways:

1. **Compaction and reduced porosity**: The rock generally becomes less porous than the precursor, and if the precursor is a sedimentary rock, the rock may become harder, such as when shale metamorphoses into slate.

2. **Change in mineralogy**: New minerals form during metamorphism as a readjustment to new conditions by a recombination of the original elements in the precursor or by **metasomatism**, the metamorphic process by which new minerals are added to the original rock, changing its composition. The new minerals can be derived from groundwater in the area or from magmatic fluids moving through the original rock body.

3. **Recrystallization**: The atoms of various elements in the rock rearrange themselves and the rock recrystallizes or becomes crystalline, crystals grow, and the rock "flows." Pressure is an important part of the cause of recrystallization, and if the pressure is not equal from all directions, sheet and chain silicate minerals will grow in the direction of least resistance, perpendicular to the direction of applied force. If the precursor is a silicate clastic, constituent grains will rotate and be deformed (squashed). If the pressure is equal from all directions, three-dimensional minerals, such as feldspars and quartz, will be favored. Lithostatic pressure occurs early as a result of subduction or due to lateral compression associated with mountain building. Hydrostatic pressure is the consequence of deep subduction.

Metamorphic rocks are classified on the basis of the mineral content of the precursor (parent rock, protolith) and/or on the kind of texture the metamorphosed rock develops. Table 2.1 lists some common metamorphic rocks, their precursors, and the metamorphic grade at which they develop.

Foliated Texture

The texture of metamorphic rocks refers to the size, shape, and arrangement of the minerals within them. When the minerals in a metamorphic rock are arranged in a nearly planar (parallel) orientation, not necessarily flat, but wavy, they have a **foliated texture**.

The following foliated textures are a result of the reaction precursors to increasing heat and temperature.

1. **Slaty cleavage** refers to the breakage characteristics of slates parallel to the alignment of fine-grained materials in a planar and parallel orientation. Slaty cleavage is a low-grade metamorphic texture. Shale, dolostone, or fine-grained rhyolite, and others, may metamorphose into **slate**.

2. **Phyllitic texture** refers to a slightly wavy to subparallel orientation of submicroscopic sheet silicate minerals, mostly very-fined-grained chlorite and/or muscovite. Rocks with phyllitic texture commonly have a pearly or satiny luster as well as a wrinkled

TABLE 2.1   Metamorphic Textures, Precursors, and Associated rocks

Texture	Precursor	Low-Grade Metamorphism	Medium-Grade Metamorphism	High-Grade Metamorphism
Nonfoliated	Shale	Hornfels	Hornfels	Hornfels
	Basalt	Hornfels	Hornfels	Hornfels
	Limestone	Altered limestone	Marble	Marble gneiss
	Dolostone	Dolostone	Marble	Marble gneiss
	Quartz Sandstone	"Quartz" sandstone	Quartzite	Quartz gneiss
	Peridotite	Serpentinite	Amphibolite	
	Conglomerate	Metaconglomerate		
	Bituminous coal	Anthracite coal		
	Limestone	*Skarn (with metasomatism)	Marble	Marble gneiss
Foliated	**Shale	Slate	Phyllite to schist	Gneiss
	Dolostone	Slate	Phyllite	Talc schist
	Basalt	Chlorite schist, biotite schist	Amphibolite	Granulite
	Gabbro	"Gabbro"	"Gabbro"	Gabbroic gneiss
	Rhyolite	Muscovite/biotite phyllite	Muscovite/biotite schist	Granitic gneiss
	Bituminous coal		Graphite schist	
	Granite	"Granite"	"Granite"	Granitic gneiss

* Skarn is the result of contact metamorphism and metasomatism of limestone.
** Shale is the only precursor that can be progressively changed into each foliated texture.

appearance. Phyllitic texture is the result of low- to medium-grade metamorphism. Slate typically metamorphoses into phyllite. Phyllites are named according to the most conspicuous mineral in the rock, such as *biotite phyllite*, *muscovite phyllite*, and so forth.

3. **Schistose texture** refers to the parallel to subparallel orientation of macroscopic sheet silicate minerals in metamorphic rocks. Biotite, muscovite, and chlorite, if present, commonly exhibit schistosity. Schistose texture is the result of medium- to high-grade metamorphism. Phyllites can metamorphose into schist. Schists are named according to the most conspicuous mineral in the rock, such as *biotite schist*, *muscovite schist*, *chlorite schist*, *hornblende schist*, and so forth.

4. **Gneissic texture** refers to alternating bands of predominantly light- and dark-colored minerals, the result of recrystallization and segregation of given minerals due to high-grade metamorphism. Lower-grade metamorphic rocks such as phyllite and schist and igneous rocks such as basalt, gabbro, or granite may metamorphose into gneiss. Gneisses are named according to the parent rock, such as *granite gneiss* or *gabbro gneiss*, and so forth.

Texture (Nonfoliated)

When the minerals in the metamorphic rock are randomly oriented, they have a **nonfoliated texture**. The precursors may have had grain sizes quite different from and commonly smaller than those of the metamorphic rock. Many nonfoliated metamorphic rocks come from precursors with simple mineralogy, and thus pressure or heat cannot rearrange the limited kinds of elements into new minerals. Rocks such as marble usually do not give much information about whether the forces involved in metamorphism were balanced (equal in all directions) or unbalanced. When the minerals in a metamorphic rock are granular or "sugary" in appearance, the nonfoliated texture is referred to as **granoblastic (crystalline)**. *Marble, quartzite,* and *hornfels* have a granoblastic texture. **Porphyroblastic texture** refers to recrystallized rocks that develop a few large minerals, commonly garnet, within a finer matrix. *Skarn* is an example of a metamorphic rock with porphyroblastic texture. Some metamorphic rocks such as *anthracite coal* have a **glassy texture**.

CLASSES OF METAMORPHISM

Classes of metamorphism are defined by the dominance of the conditions producing the change (metamorphism). **Contact metamorphism** refers to the alteration of rock adjacent to and surrounding a magma intrusion where heat and sometimes fluids from a "wet" magma or superheated groundwater play a major role in the metamorphism. The heat and fluids (if any) produce a zone of "baked" (dry conditions) and/or otherwise altered (wet conditions) rock adjacent to the magma (Figure 2.4). This zone is called the **contact aureole** or **halo**. It is confined to the immediate area of the emplaced (intruded) igneous body responsible for the heat. In Figure 2.4, the contact aureole extends furthest into the sandstone because the porosity and permeability of the sandstone allows greater lateral migration of hot fluids through the rock. Immediately adjacent to the intrusion, hornfels, a contact metamorphic rock, may form from shales, even under dry conditions. Garnet and pyroxene, amphibole, epidote, chlorite, and serpentine (in order of increasing distance

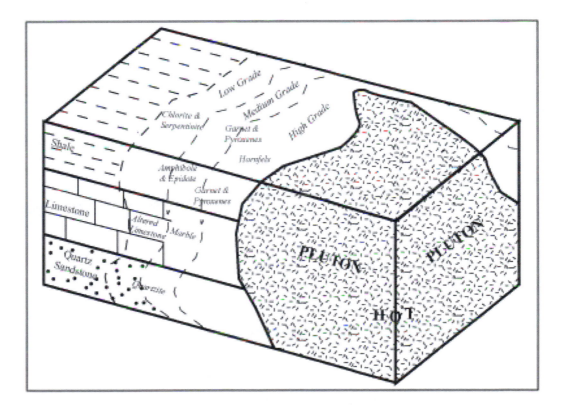

FIGURE 2.4 Contact Metamorphism.

from the heat source) are mineral phases that may form in shale. Limestone will meta-morphose into marble at medium-grade metamorphic conditions and melt at the higher temperature nearest the intrusion. Quartz sandstone metamorphoses into quartzite at medium-grade metamorphism and will also melt at the higher temperature nearest the intrusion.

Regional metamorphism is the large-scale alteration of the parent (precursor or "country") rock of an area subjected to tectonic pressures (plate collisions) and heat (usu-ally the result of subduction, deep burial). The characteristic silicate mineral groups, and ultimately feldspars, reflect the variations in conditions of the metamorphism of huge volumes of rocks.

GRADES OF METAMORPHISM

The metamorphic changes in precursors reflect the grade of metamorphism and the pro-gressive sequence of imposed pressure and temperature conditions they were subjected to. The types and grades of regional metamorphism (Figure 2.5) are indicated by what is called metamorphic facies. **Metamorphic facies** are the assemblages of minerals that form as a result of different conditions of metamorphism.

1. **Very-low-grade metamorphism** takes place under conditions of low temperature and high lithostatic pressure (pressure imposed by the weight of the overlying mate-rial), and results in the formation of minerals that make up what is called the **zeolite and prenhite-pumpellyite facies** (Figure 2.5). The minerals of this metamorphic fa-cies tend to be laumonite or wairakite (both zeolites), or prenhite, pumpellyite, albite, calcite, and quartz.

2. **Low-grade metamorphism** takes place under conditions of low temperature (200−450°C) and low lithostatic pressure and can result in the progression of shale (sedimentary rock) to slate a metamorphic rock with slaty cleavage (textural term). Clay minerals are sheet silicates, and in shale the clay minerals are very small and

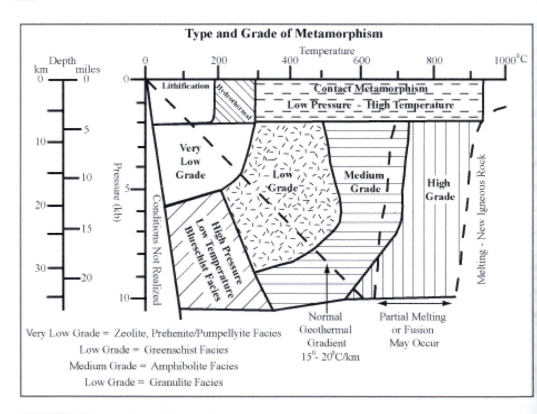

FIGURE 2.5 Metamorphic Grades.

randomly oriented relative to one another. The change to slate involves growth (enlargement) of the clays as they recrystallize, and rotate into a more parallel alignment. The **greenschist facies** is indicative of low-grade metamorphism and is marked by the formation and stability of greenish minerals such as chlorite (sheet silicate) and epidote.

3. **Medium-grade metamorphism** occurs at temperatures from about 400−700°C. At this grade, slate can metamorphose into a phyllite (textural term), and as the temperature and lithostatic pressure increase, phyllite can change into a schist (textural term). The change is a continuation of the recrystallization discussed for the change from shale to slate. The facies associated with medium-grade metamorphism is known as the **amphibolite facies** because of the amount of stable amphiboles (hornblende) commonly present as well as the minerals garnet, biotite, and staurolite.

4. **High-grade metamorphism,** at approximately 700−900°C and very high hydrostatic pressures, can result in the formation of gneiss (textural term) from schist. Rocks formed at high-grade metamorphism belong to the granulite facies. **Granulite** is a textural term for rocks with equigranular crystals. Minerals such as hornblende, pyroxene, and sillimanite are common minerals stable at the high temperatures and pressures of high-grade metamorphism.

Metamorphic rock pictures and other information can be found on the Earth and Space Science website (http://ess.lamar.edu). Click on the People tab, Staff, Karen M. Woods, Teaching, Physical Geology Lab, Metamorphic Rocks.

Key to the Identification of Common Metamorphic Rocks

FOLIATED TEXTURE
(Recrystallized crystals are arranged in an orderly fashion)

Slaty Cleavage Texture → Foliation Microscopic → **SLATE**

Phyllitic Texture → Foliation Submicroscopic → **PHYLLITE**

Schistose Texture → Foliation Macroscopic → **SCHIST**

Gneissic Texture → Foliation recrystallized minerals, segregated into distinct layers → **GNEISS**

FOLIATED TEXTURE
(Recrystallized crystals are not arranged in an orderly fashion)

Granoblastic Texture
→ Entire rock reacts to acid, "fizzes" → **MARBLE**
→ "Sugary" look, rock does not react to acid → **QUARTZITE**
→ Dark colored, may have spots → **HORNFELS**

Porphyroblastic Texture
→ Metamorphosed conglomerate → **METACONGLOMERATE**
→ Large minerals (garnets) in finer matrix → **SKARN**

Glassy Texture → Lightweight, black, conchoidal fracture → **ANTHRACITE COAL**

Summary of Metamorphic Rocks and Their Precursors

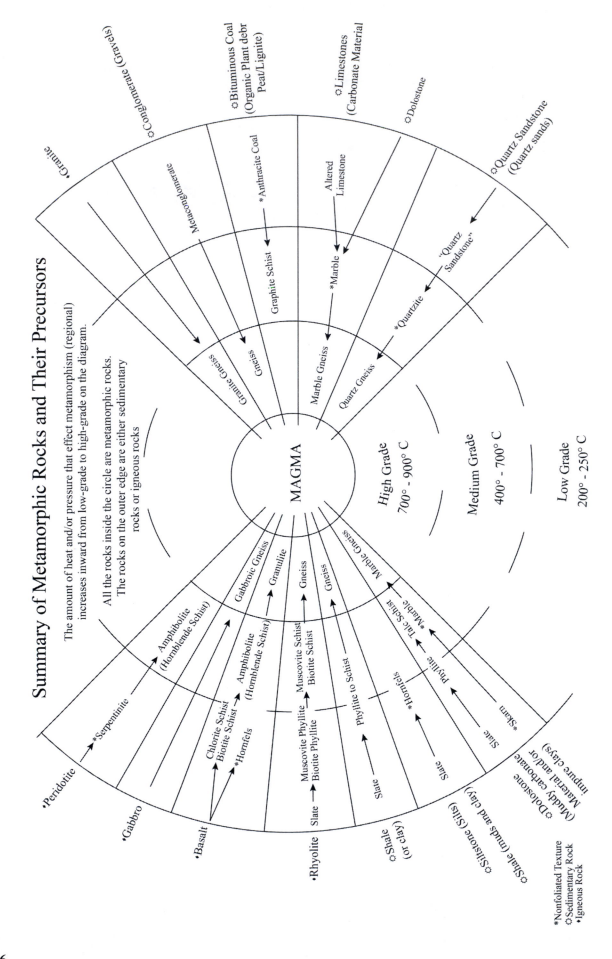

The amount of heat and/or pressure that effect metamorphism (regional) increases inward from low-grade to high-grade on the diagram.

All the rocks inside the circle are metamorphic rocks. The rocks on the outer edge are either sedimentary rocks or igneous rocks

MAGMA

High Grade
700° - 900° C

Medium Grade
400° - 700° C

Low Grade
200° - 250° C

*Nonfoliated Texture
◊Sedimentary Rock
•Igneous Rock

EXERCISE 2.3: METAMORPHIC
ROCK IDENTIFICATION

Fill in all the spaces in the exercise. DO NOT leave any blanks. If the answer is not applicable, write N/A in the blank.

The identification of unknown metamorphic rocks is determined by the visual aspects of each sample.

Texture:

Study each rock sample and determine the general arrangement of the constituent minerals within the rock.

If the minerals appear to be randomly arranged, the texture is nonfoliated.

Record the correct descriptive nonfoliated texture for the rock:

granoblastic or crystalline, porphryoblastic, or glassy.

If the minerals appear to be arranged in an orderly fashion, the texture is foliated.

Record the correct descriptive foliated texture for the rock:

slaty cleavage, schistose, phyllitic, or gneissic.

Recognizable Minerals:

List any recognizable minerals within each rock.

Precursors:

List the precursor(s) for each rock.

For each rock, list an igneous precursor, a sedimentary precursor, and a metamorphic precursor, if it applies. Write N/A if a precursor does not exist. Some rocks will only have one precursor; others may have multiple possible precursors.

Description:

Describe the rock. Note any special properties in this section.

Properties / Rock Name	Texture (be specific)	Recognizeable Minerals	Precursors Igneous, Sedimentary, & Metamorphic	Description
			Ig: Sed: Meta:	
			Ig: Sed: Meta:	
			Ig: Sed: Meta:	
			Ig: Sed: Meta:	
			Ig: Sed: Meta:	
			Ig: Sed: Meta:	

Properties / Rock Name	Texture (be specific)	Recognizeable Minerals	Precursors Igneous, Sedimentary, & Metamorphic			Description
			Ig:	Sed:	Meta:	
			Ig:	Sed:	Meta:	
			Ig:	Sed:	Meta:	
			Ig:	Sed:	Meta:	
			Ig:	Sed:	Meta:	
			Ig:	Sed:	Meta:	

Properties / Rock Name	Texture (be specific)	Recognizeable Minerals	Precursors Igneous, Sedimentary, & Metamorphic			Description
			Ig:	Sed:	Meta:	
			Ig:	Sed:	Meta:	
			Ig:	Sed:	Meta:	
			Ig:	Sed:	Meta:	
			Ig:	Sed:	Meta:	
			Ig:	Sed:	Meta:	

Properties / Rock Name	Texture (be specific)	Recognizeable Minerals	Precursors Igneous, Sedimentary, & Metamorphic			Description
			Ig:	Sed:	Meta:	
			Ig:	Sed:	Meta:	
			Ig:	Sed:	Meta:	
			Ig:	Sed:	Meta:	
			Ig:	Sed:	Meta:	
			Ig:	Sed:	Meta:	

ROCK PROPERTY LIST

Amphibolite (hornblende schist)—Amphibolite is the result of metamorphism of fine-grained mud or shale (sedimentary), basalt, gabbro (igneous), or lower-grade metamorphic rock such as slate or hornfels, which changed the mineralogy of the precursor to predominantly hornblende. Amphibolite has a foliated-schistose texture. See **schist** for general characteristics.

Andesite—Andesite is a light-gray, red, or pink extrusive (volcanic) rock. Andesite has an intermediate composition and generally has an aphanitic or porphyritic texture. Andesite is the finer-grained intermediate rock composed primarily of plagioclase feldspar with lesser amounts of biotite, hornblende, and pyroxene (augite).

Basalt—Basalt is a dark-gray to black, extrusive igneous (volcanic) rock. Basalt has a mafic composition and aphanitic texture.

> **Porphyritic Basalt**—Porphyritic basalt has a porphyritic texture, a fine-grained aphanitic matrix (dark color) in which lighter-colored phenocrysts (larger crystals, commonly olivine) are embedded. Porphyritic basalt is the result of, first, slow cooling of magma, followed by rapid cooling of the extruded lava.

Breccia—Breccia is a clastic, coarse-grained, sedimentary rock and has a terrigenous mode of origin. Breccia is composed of angular rock fragments (> 2 mm in size) that vary according to the nature of the source material (igneous, sedimentary). The larger clasts are contained within a finer-grained matrix. The angular clasts indicate that the source was either composed of hard, resistant rock or that the clasts were not transported far from the source.

Chalk—Chalk is a biogenic sedimentary rock resulting from the accumulation of microscopic hard parts of calcareous marine organisms. Chalk has a bioclastic-fine-grained texture, is white to gray, and reacts strongly to dilute HCl acid. The "White Cliffs of Dover" in England are composed of chalk.

Chert—Chert is a sedimentary precipitate that occurs in a variety of colors. Chert is a cryptocrystalline (texture) precipitate of silicon dioxide (SiO_2) and has conchoidal fracture. Chert develops in many ways in several environments, including marine and lacustrine, and is commonly found as nodules in chalk deposits.

Coal—Coal is a sedimentary rock that forms from the accumulation of plant material in lacustrine, swampy, lagoonal, or deltaic environments. Coal has a biogenic origin and a bioclastic-coarse-grained texture. Coal is a "fossil fuel," burned to generate electricity.

> **Anthracite coal**—Anthracite is metamorphosed bituminous coal and is the result of low-grade regional metamorphism. Anthracite has a nonfoliated, glassy, almost metallic texture, and conchoidal fracture. Anthracite resembles obsidian, but is softer and is not as heavy.
>
> **Bituminous Coal**—Bituminous coal is a dull black sedimentary rock with evidence of the sedimentary stratification that is characteristic of sedimentary rocks. It has a bioclastic, coarse-grained texture, although the texture may not be easily recognized as such in hand specimens.
>
> **Lignite**—Lignite is the brownish-black sedimentary precursor of bituminous coal and is sometimes referred to as "brown coal."
>
> **Peat**—Peat is sediment (not a coal) composed of shreds of plant tissue. Peat is the precursor to lignite.

Conglomerate—Conglomerate is a clastic, coarse-grained, sedimentary rock and has a terrigenous origin. Conglomerate is composed of rounded rock fragments (>2 mm in size) that vary according to the nature of the source material. The larger clasts are contained within a finer-grained matrix. Conglomerate is indicative of fluvial or near-shore marine environments. The cement that holds the rock clasts together can be silica, calcite, or iron oxide. Conglomerate is similar in appearance to concrete, a man-made equivalent.

Coquina—Coquina is a <u>biogenic</u>, <u>bioclastic</u>, <u>coarse-grained</u>, <u>sedimentary</u> rock. Coquina is a variety of limestone composed of broken pieces of shell material (shell hash). Coquina reacts strongly to dilute HCl acid, and hand specimens resemble granola bars.

Diorite—Diorite is light to dark gray and has a <u>phaneritic</u> or <u>porphyritic</u> texture. The composition of diorite is <u>intermediate</u>. Diorites are <u>intrusive</u> (plutonic) <u>igneous</u> rocks that are composed primarily of plagioclase feldspar with lesser amounts of biotite, hornblende, and pyroxene (augite).

Dolostone—Dolostone is a marine <u>chemical</u> <u>precipitate</u> and has a <u>crystalline</u> texture. Most dolostones will react to dilute HCl if it is powdered.

Fossiliferous Limestone—Fossiliferous limestone has a <u>biogenic</u> origin and is a <u>bioclastic</u>, <u>coarse</u>-grained, <u>sedimentary</u> rock. Fossiliferous limestone is composed of fine-grained lime (calcareous) mud that contains large visible fossils, and is generally light in color. The rock usually develops in marine environments and reacts strongly with dilute HCl acid.

Gabbro—Gabbro is a dark green-gray to blue-black, <u>phaneritic</u> (medium-grained), <u>intrusive</u> (plutonic), <u>igneous</u> rock and is <u>mafic</u> in composition.

Granite—Granite is a light-colored, <u>intrusive</u> (plutonic), <u>igneous</u> rock with a <u>felsic</u> composition and a <u>phaneritic</u> texture. Granite is composed of orthoclase feldspar, quartz, mica, amphibole, and/or low-temperature plagioclase feldspar.

> **Pink Granite**—Pink granite owes its color to the presence of "pink" orthoclase.

> **White granite**—White granite owes its color to the presence of white feldspars (orthoclase and/or microcline).

Granite Gneiss—Gneiss is a foliated textural term used to describe <u>metamorphic</u> rocks with alternating bands of light and dark minerals. Gneiss develops as the result of <u>high</u>-grade <u>regional</u> <u>metamorphism</u> of precursor rocks. The name is modified on the basis of the actual precursor, if known (e.g., granite gneiss, gabbro gneiss, etc.).

Hornfels—Hornfels (hornfelses, plural) is a <u>contact</u> <u>metamorphic</u> rock with a <u>granoblastic</u>, <u>nonfoliated</u> texture. The precurser to hornfels is clay, shale, or basalt. Hornfels is light to dark gray in color and may display spots.

Llanite—See rhyolite.

Marble—Marble is a <u>metamorphic</u> rock with a <u>granular</u> or <u>crystalline</u>, <u>nonfoliated</u> texture. Marble develops as the result of <u>medium</u>-grade <u>regional</u> or <u>contact</u> metamorphism of a limestone or dolostone precursor. Marble can be any color and resembles quartzite, but is softer and will react strongly to dilute HCl acid.

Metaconglomerate—Metaconglomerate is a <u>metamorphic</u> rock with a <u>granoblastic</u> or <u>crystalline</u>, <u>nonfoliated</u> texture developed as the result of <u>very</u>-<u>low</u>-grade to <u>low</u>-grade <u>regional</u> metamorphism. The precursor of metaconglomerate is conglomerate.

Micrite—Micrite is a <u>biogenic</u> <u>sedimentary</u> rock with a <u>bioclastic</u>, <u>very</u>-<u>fine</u>-grained to <u>fine</u>-grained or <u>cryptocrystalline</u> texture, and conchoidal fracture. Micrite tends to resemble chert, but is softer and can be easily distinguished by its reaction to dilute HCl acid. Micrite results from compaction of fine lime mud, in marine environments.

Obsidian—Obsidian is an <u>extrusive</u> <u>igneous</u> rock. Obsidian has a <u>glassy</u> texture due to rapid cooling, and conchoidal fracture. Most obsidian is <u>felsic</u> in composition, but <u>intermediate</u> <u>to</u> <u>mafic</u> obsidian is known.

> **Pumice**—Pumice is "puffed" obsidian with a microscopic <u>vesicular</u> texture. Pumice is derived from the explosive release of vapors and gases from erupted felsic (viscous) lava that creates very small vesicles (holes) between glass fibers. Rhyolites are commonly pumiceous. Pumice is light in color and so lightweight that it will float on water. The composition is <u>felsic</u> or <u>intermediate</u>. See obsidian above for other properties.

> **Scoria**—Scoria is a variety of <u>mafic</u> obsidian with a <u>vesicular</u> texture. Scoria is usually a dark gray to black or reddish brown. Scoria is formed when water and gases bubble up through basaltic or coarser porphyritic lava and form holes

(vesicles). Scoria is derived from "puffed up" basalt or porphyritic basalt and has a mafic composition. See obsidian above for other properties.

Peridotite—Peridotite is an ultramafic, intrusive igneous rock with a phaneritic texture. Peridotite that is composed essentially of olivine is called dunite.

Phyllite—Phyllite is a name given to metamorphic rocks that have a phyllitic texture. Phyllite is the result of low-grade regional metamorphism of clay and shales (sedimentary), tuff and rhyolite (igneous), or slate (metamorphic). Phyllites are composed of predominantly micaceous minerals that give a satiny sheen to the rock, and they often have a wrinkled appearance.

Pumice—See obsidian.

Quartzite—Quartzite is a metamorphic rock that forms as the result of contact or medium-grade regional metamorphism of quartz sandstone (precursor). Quartzite has a nonfoliated, crystalline or granular "sugary" texture. Quartzite can be any light color, and resembles marble, but is much harder and will not react to dilute HCl acid.

Rhyolite—Rhyolite is an extrusive (volcanic) igneous rock. Rhyolite is of felsic composition and has an aphanitic or finely porphyritic texture. Rhyolite ranges from light gray, to pink, to red, or brown.

> **Llanite (Melarhyolite)**—Llanite is a dark-colored porphyritic rhyolite. Llanite is composed of phenocrysts of quartz, anorthoclase, albite, and microcline in an aphanitic matrix of quartz, feldspar, biotite, amphibole, and opaques. The dark color of the matrix is the result of the presence of 15–30% biotite. The quartz phenocrysts are blue due to the inclusion of needle-like rutile (sagenite variety).

Rock Gypsum—Rock gypsum is a sedimentary rock with an evaporitic origin. Rock gypsum is composed almost entirely of the mineral gypsum ($CaSO_4.2H_2O$). Rock gypsum has a fine to coarse, crystalline texture, varies in color (usually light), and commonly has alternating bands of light and dark impurities (muds). Rock gypsum is formed in marine or lacustrine environments, and is fingernail-soft.

Rock Salt—Rock salt is a sedimentary rock with an evaporitic origin. Rock salt is composed almost entirely of the mineral halite (NaCl). Rock salt has a crystalline texture, and can be transparent or stained various colors by impurities. Rock salt is developed in marine or lacustrine environments and has a salty taste. *(Tasting of laboratory specimens is not recommended.)*

Sandstone—Sandstone is a sedimentary rock with a clastic, medium-grained texture and a terrigenous origin. Sandstone is composed mainly of quartz sand grains. The rock may vary in color, as determined by the cementing agents, impurities. Sandstone is associated with marine, fluvial, or aeolian depositional environments.

Schist—Schist is a term used to describe metamorphic rocks that have a parallel to sub-parallel orientation of macroscopic sheet silicate, micaceous minerals. Schist is developed as a result of medium- to high-grade regional metamorphism of precursor rocks. The name is modified on the basis of the most predominantly appearing mineral(s) in each rock (e.g., biotite schist, chlorite schist, etc.).

> **Chlorite Schist**—Chlorite schist is the result of the metamorphism of fine-grained mud or shale (sedimentary), basalt (igneous), or lower-grade metamorphic rock such as slate, which changed the mineralogy of the precursor to predominantly chlorite. Chlorite schist has a foliated-schistose texture and a greenish, platy appearance. See above for general characteristics.

> **Hornblende Schist**—See amphibolite.

> **Muscovite Schist**—Muscovite schist is the result of metamorphism of fine-grained mud or shale (sedimentary), rhyolite (igneous), or lower-grade metamorphic rock such as slate, which changed the mineralogy to predominantly muscovite. Muscovite schist has a foliated-schistose texture and resembles folded, broken, and "glued" together flakes of muscovite. See schist above for general characteristics.

Scoria—See obsidian.

Shale—Shale is a <u>sedimentary</u> rock with a <u>terrigenous</u> origin. Shale has a <u>clastic</u>, <u>very-fine-grained</u> (< 1/256 mm) texture. Shale is associated with marine and lacustrine environments. Shale looks similar to slate, but is softer, and sounds dull when tapped. Shale can be scratched easily with a penny.

Skarn—Skarn is formed by the <u>contact</u> <u>metamorphism</u> and metasomatism of limestone or dolomite (carbonate) precursors. Skarn has a <u>nonfoliated</u>, <u>porphyroblastic</u> texture. Skarn varieties are based on the predominant minerals in the rock.

Slate—Slate is a <u>metamorphic</u> rock that forms as the result of the <u>regional</u>, <u>low-grade</u> metamorphism of clay, shale, basalt, rhyolite or other fine-grained rock (precursors). Slate has a <u>foliated</u>, <u>slaty cleavage</u> texture. Slate is harder than shale. When tapped, slate has a higher-pitched ping and, more often than not, <u>cannot</u> be scratched by a penny.

Syenite—Syenite is an <u>intrusive</u> (plutonic) <u>igneous</u> rock with a <u>felsic</u> composition and <u>phaneritic</u> texture. Syenite resembles granite, except it contains little or no quartz, and may have a greenish cast.

USES FOR COMMON ROCKS

Igneous Rocks

Gabbro—Polished slabs of gabbro can be used as an interior decorative stone, to face walls, floors, or counter tops.

Granite—Because of its felsic (sialic) composition and coarse, phaneritic texture, granite is an excellent exterior building stone. Polished slabs of it are used to face buildings, or for interior rockwork such as floor tiles, counter tops, and so forth. Granite is often used for tombstones because of its durability.

Obsidian—Obsidian was cherished by stone-age people, such as American Indians, for the fabrication of knives, scrapers, drills, engravers, projectile points, and other tools, because the rock can be shaped (has conchoidal fracture) and because it is hard. Obsidian serves modern medicine as extremely sharp surgical knives that create a cleaner incision than steel scalpels.

Pumice—Pumice is used as an abrasive in hand soap, a foot polisher (smoother), and as material for sculptures.

Peridotite—Gem-quality peridotite is used as jewelry (peridot).

Sedimentary Rocks

Bituminous coal—Bituminous coal is a medium-grade coal (50–80% carbon) used for fuel.

Chert—Chert has historically been used to make arrowheads. It is also used in jewelry and as decorative material.

Lignite—Lignite is a low-grade brownish coal used as fuel.

Limestone—Some types of limestone are used as decorative stone in homes and businesses.

Peat—Peat is composed of plant remains and is used as fuel when abundant (Wicander & Monroe, 2002).

Rock Salt—Rock salt is used to melt ice and snow on roadways. Pure rock salt (halite) is used in food preparation, as a preservative, and as a flavor enhancer.

Metamorphic Rocks

Anthracite coal—Anthracite coal is a high-grade coal (greater than 90% carbon) used for fuel.

Marble—Marble is used for decorative floor tiles, wall tiles, and counter tops.

Quartzite—Quartzite is used as construction material.

Slate—Slate is used for chalkboards and as roofing material.

ALABASTER
GYPSUM

$CaSO_4 \cdot n2H_2O$

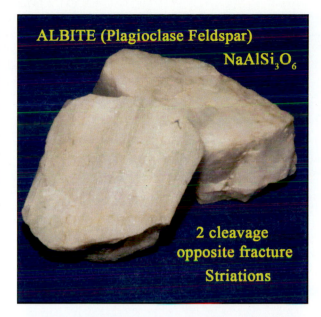

ALBITE (Plagioclase Feldspar)

$NaAlSi_3O_6$

2 cleavage
opposite fracture
Striations

APATITE

$Ca_5(PO_4)_3(F,Cl,OH)$

AUGITE

2 cleavages,
fracture
opposite

Short Prismatic
Crystals

$(Ca,Na)(Mg,Fe,Al)(Al,Si)_2O_6$

BAUXITE
$AlO(OH)$

"Mineraloid"

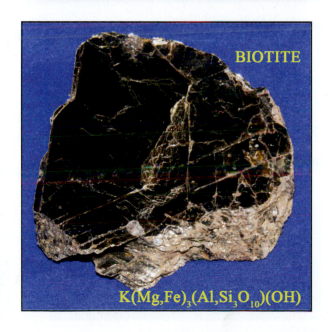

BIOTITE

$K(Mg,Fe)_3(Al,Si_3O_{10})(OH)$

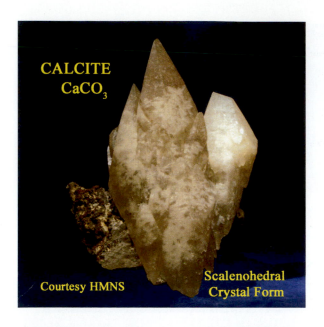

CALCITE
CaCO$_3$

Courtesy HMNS

Scalenohedral
Crystal Form

CALCITE
CaCO$_3$

3 Cleavages,
6 planes

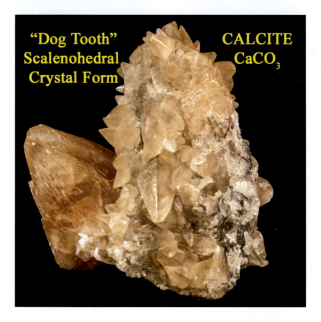

"Dog Tooth"
Scalenohedral
Crystal Form

CALCITE
CaCO$_3$

CALCITE
CaCO$_3$

"Nailhead
Spar"

CHLORITE

$(Mg,Fe)_3(Si,Al)_4O_{10}(OH)_2 \cdot (Mg,Fe)_3(OH)_6$

CORUNDUM

Al$_2$O$_3$

Hexagonal Crystal Form

FLUORITE
CaF$_2$

Cubic
Crystal Form
Octahedral Cleavage

Courtesy HMNS

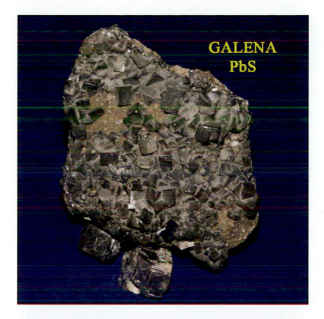

GALENA
PbS

GARNET

Massive
Variety

Dodecahedral
Crystal
Form

Complex Fe, Mg, Mn,
Ca, Al silicate

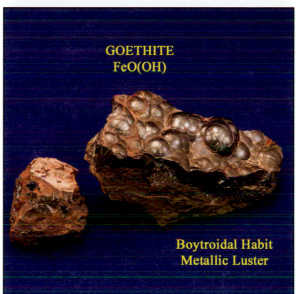

GOETHITE
FeO(OH)

Boytroidal Habit
Metallic Luster

GRAPHITE
C

SATIN SPAR GYPSUM
CaSO$_4 \cdot n$2H$_2$O

SELENITE GYPSUM
$CaSO_4 \cdot n2H_2O$

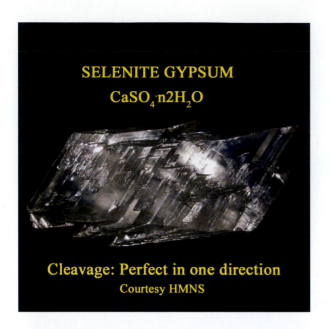

Cleavage: Perfect in one direction
Courtesy HMNS

HALITE
NaCl

Courtesy HMNS

**Cubic
Crystal Form**

HALITE
NaCl

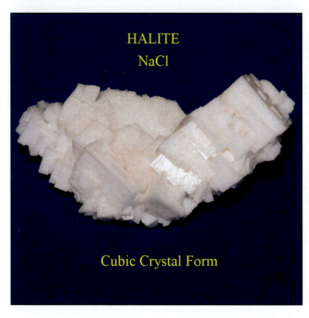

Cubic Crystal Form

OOLITIC HEMATITE

Fe_2O_3

SPECULAR HEMATITE

Fe_2O_3

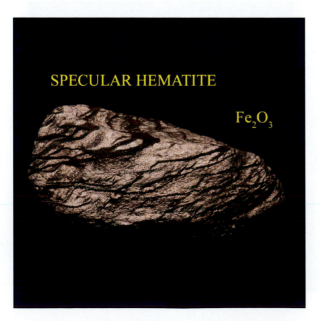

HORNBLENDE

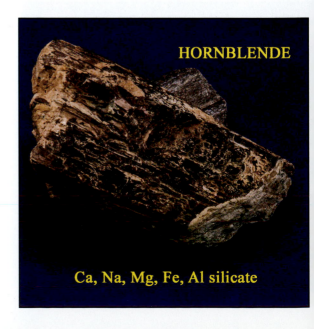

Ca, Na, Mg, Fe, Al silicate

KAOLINITE

$KAl_4Si_4O_{10}(OH)_8$

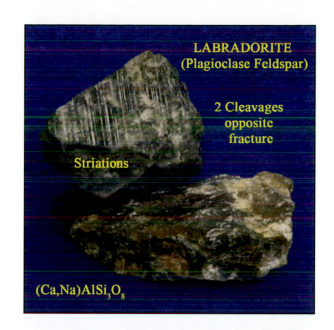

LABRADORITE
(Plagioclase Feldspar)

2 Cleavages
opposite
fracture

Striations

$(Ca,Na)AlSi_3O_8$

MAGNETITE

Fe_3O_4

Octahedral
Crystal Form

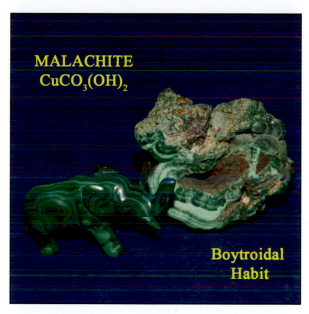

MALACHITE
$CuCO_3(OH)_2$

Boytroidal
Habit

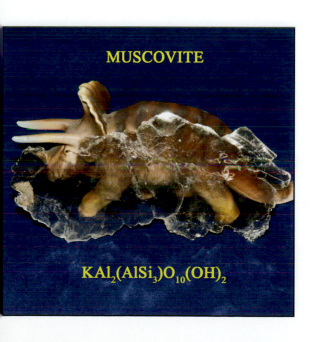

MUSCOVITE

$KAl_2(AlSi_3)O_{10}(OH)_2$

OLIVINE

$(Mg,Fe)Si_2O_4$

ORTHOCLASE FELDSPAR

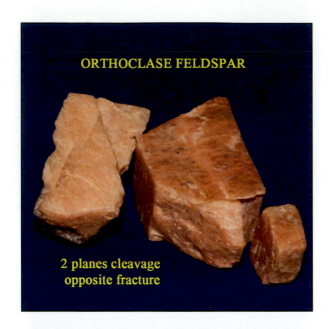

2 planes cleavage
opposite fracture

PYRITE
FeS_2

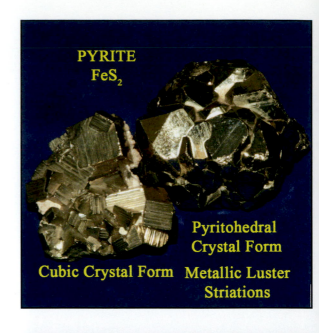

Pyritohedral
Crystal Form

Cubic Crystal Form   Metallic Luster
Striations

MILKY QUARTZ

SiO_2

ROCK CRYSTAL QUARTZ

SiO_2

Hexagonal Crystal Form

ROSE QUARTZ
SiO_2

SMOKY QUARTZ
SiO_2

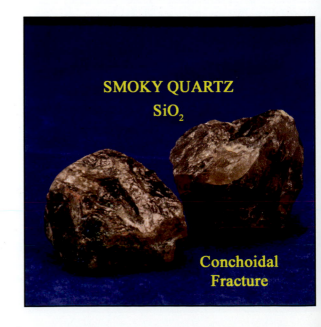

Conchoidal
Fracture

SPHALERITE
ZnS

SULFUR
S

TALC

$Mg_3Si_4O_{10}(OH)_2$

ANDESITE

Porphyritic
Texture

Composition: Intermediate

AMYGDALOIDAL BASALT

Amygdaloidal Texture   Mafic Composition

BASALT

Mafic Composition   Aphanitic Texture

OLIVINE BASALT PORPHYRY

Mafic
Composition

Porphyritic
Texture

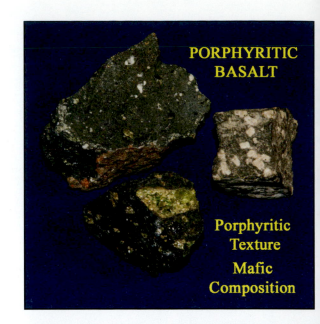

**PORPHYRITIC
BASALT**

Porphyritic
Texture

Mafic
Composition

DIORITE

Phaneritic Texture   Intermediate Composition

GABBRO

Phaneritic Texture   Mafic Composition

GRANITE

Phaneritic
Texture

Porphyritic
Texture

Felsic
Composition

**LLANITE
(Melarhyolite)**

Porphyritic Texture

Felsic Composition

OBSIDIAN

Glassy Texture

Composition: Felsic, Intermediate or Mafic

ORTHOCLASE-QUARTZ PEGMATITE

Pegmatitic Texture

Felsic Composition

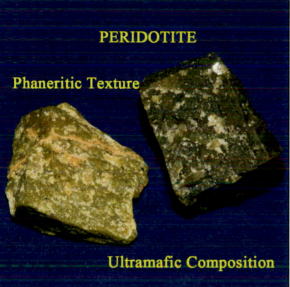

PERIDOTITE

Phaneritic Texture

Ultramafic Composition

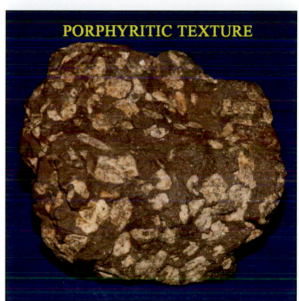

PORPHYRITIC TEXTURE

PUMICE

Vesicular Texture   Felsic Composition
(Usually)

RHYOLITE

Felsic Composition

Aphanitic or Porphyritic Texture

SCORIA

Vesicular Texture   Mafic Composition
(Usually)

VITROPHYRE

Porphyritic Texture

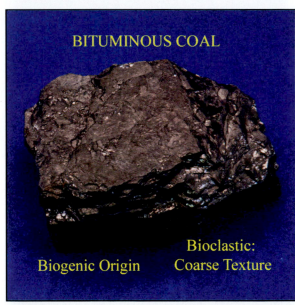

BITUMINOUS COAL

Biogenic Origin   Bioclastic:
Coarse Texture

BRECCIA

Angular Pebbles

Bioclastic: Coarse Texture
Terrigenous Origin

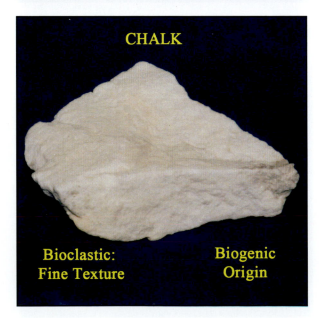

CHALK

Bioclastic:
Fine Texture   Biogenic
Origin

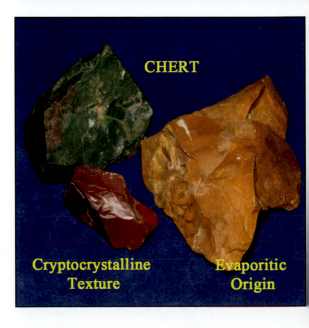

CHERT

Cryptocrystalline
Texture   Evaporitic
Origin

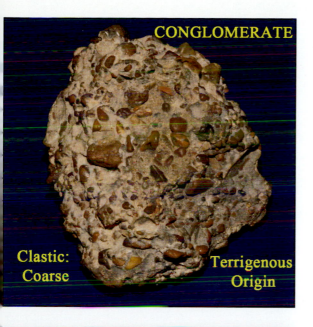

CONGLOMERATE

Clastic: Coarse

Terrigenous Origin

COQUINA

Biogenic Origin
Bioclastic: Medium-Coarse Texture

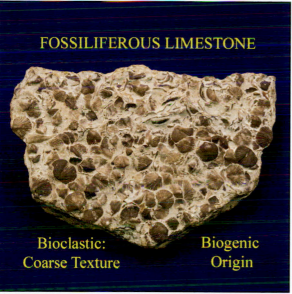

FOSSILIFEROUS LIMESTONE

Bioclastic: Coarse Texture

Biogenic Origin

LIGNITE

Biogenic Origin
Bioclastic: Coarse Texture

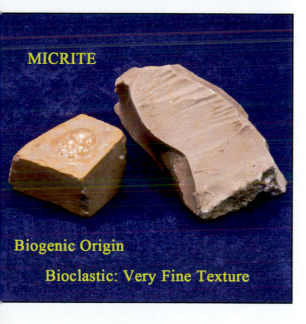

MICRITE

Biogenic Origin

Bioclastic: Very Fine Texture

PEAT

Biogenic Origin
Bioclastic: Coarse Texture

ROCK GYPSUM
Evaporitic Origin
Crystalline Texture

ROCK SALT
Crystalline Texture
Evaporitic Origin

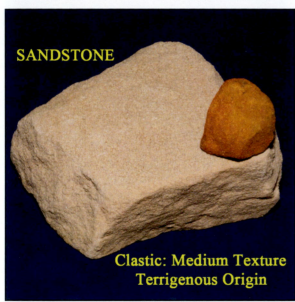

SANDSTONE
Clastic: Medium Texture
Terrigenous Origin

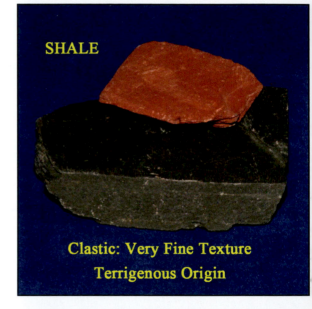

SHALE
Clastic: Very Fine Texture
Terrigenous Origin

ANTHRACITE COAL
Glassy Texture
Biogenic Origin

CHLORITE SCHIST
Schistose Texture

GRANITE GNEISS

Gneissic Texture

HORNBLENDE SCHIST

Schistose Texture

"SPOTTED" HORNFELS

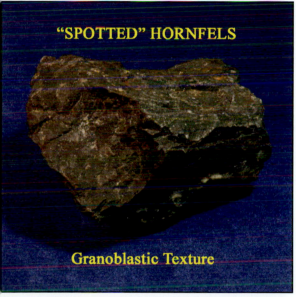

Granoblastic Texture

MARBLE

Crystalline Texture

METACONGLOMERATE

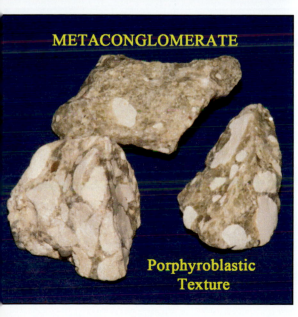

Porphyroblastic Texture

MUSCOVITE SCHIST

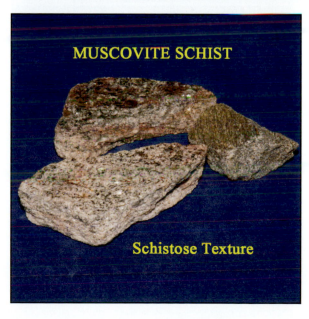

Schistose Texture

PHYLLITE

Phyllitic
Texture

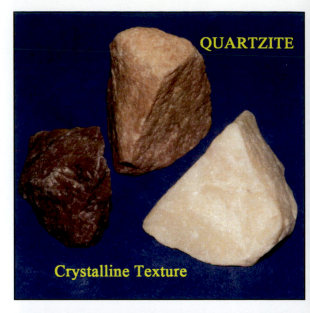

QUARTZITE

Crystalline Texture

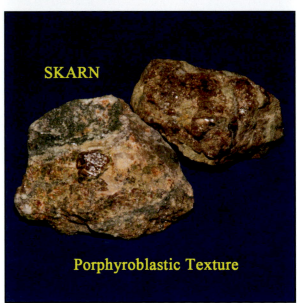

SKARN

Porphyroblastic Texture

SLATE

Slaty Cleavage Texture

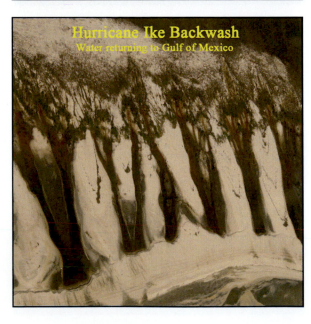

Hurricane Ike Backwash
Water returning to Gulf of Mexico

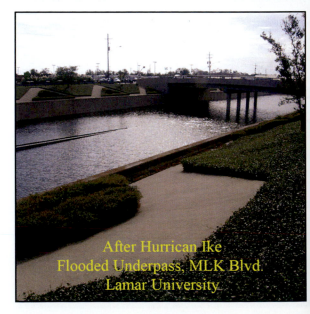

After Hurrican Ike
Flooded Underpass, MLK Blvd.
Lamar University

Topographic Map Symbols

....
nt ....
equivalent ....
r equivalent ....
r monument ....

STEMS

rvey System:
ge line
ful ....
....
ful ....
rner; found closing corner ....
meander corner .... WC MC

ge line ....

ng claim; monument ....

TED FEATURES

....

LATED FEATURES

median strip ....
uction ....
ss ....

ELATED FEATURES

f employment: small; large ...
etc.: small; large
t
ater); windmill
large
arge

c area
rge

RAILROADS AND RELATED FEATURES

Standard gauge single track; station ....
Standard gauge multiple track ....
Abandoned ....
Under construction ....
Narrow gauge single track ....
Narrow gauge multiple track ....
Railroad in street ....
Juxtaposition ....
Roundhouse and turntable ....

TRANSMISSION LINES AND PIPELINES

Power transmission line: pole; tower ....
Telephone or telegraph line ....
Aboveground oil or gas pipeline ....
Underground oil or gas pipeline ....

CONTOURS

Topographic:
    Intermediate ....
    Index ....
    Supplementary ....
    Depression ....
    Cut; fill ....

Bathymetric:
    Intermediate ....
    Index ....
    Primary ....
    Index Primary ....
    Supplementary ....

MINES AND CAVES

Quarry or open pit mine ....
Gravel, sand, clay, or borrow pit ....
Mine tunnel or cave entrance ....
Prospect; mine shaft ....
Mine dump ....
Tailings ....

SURFACE FEATURES

Levee ....
Sand or mud area, dunes, or shifting sand ......
Intricate surface area ....
Gravel beach or glacial moraine ....
Tailings pond ....

VEGETATION

Woods ....
Scrub ....
Orchard ....
Vineyard ....
Mangrove ....

COASTAL FEATURES

Foreshore flat ....
Rock or coral reef ....
Rock bare or awash ....
Group of rocks bare or awash ....
Exposed wreck ....
Depth curve; sounding ....
Breakwater, pier, jetty, or wharf ....
Seawall ....

BATHYMETRIC FEATURES

Area exposed at mean low tide; sounding datum .
Channel ....
Offshore oil or gas: well; platform ....
Sunken rock ....

RIVERS, LAKES, AND CANALS

Intermittent stream ....
Intermittent river ....
Disappearing stream ....
Perennial stream ....
Perennial river ....
Small falls; small rapids ....
Large falls; large rapids ....

Masonry dam ....

Dam with lock ....

Dam carrying road ....

Intermittent lake or pond ....
Dry lake ....
Narrow wash ....
Wide wash ....
Canal, flume, or aqueduct with lock ....
Elevated aqueduct, flume, or conduit ....
Aqueduct tunnel ....
Water well; spring or seep ....

GLACIERS AND PERMANENT SNOWFIELDS

Contours and limits ....
Form lines ....

SUBMERGED AREAS AND BOGS

Marsh or swamp ....
Submerged marsh or swamp ....
Wooded marsh or swamp ....
Submerged wooded marsh or swamp ....
Rice field ....
Land subject to inundation ....

Cem

MILE SCALE 1:62 500

```
1          0          1          2          3          4          5 MILES
```

UNITED STATES
DEPARTMENT OF THE INTERIOR
GEOLOGICAL SURVEY

TOPOGRAPHIC
MAP INFORMATION AND SYMBOLS
MARCH 1978

QUADRANGLE MAPS AND SERIES

Quadrangle maps cover four-sided areas bounded by parallels of latitude and meridians of longitude. Quadrangle size is given in minutes or degrees.

Map series are groups of maps that conform to established specifications for size, scale, content, and other elements.

Map scale is the relationship between distance on a map and the corresponding distance on the ground.

Map scale is expressed as a numerical ratio and shown graphically by bar scales marked in feet, miles, and kilometers.

NATIONAL TOPOGRAPHIC MAPS

Series	Scale	1 inch represents	1 centimeter represents	Standard quadrangle size (latitude-longitude)	Quadrangle area (square miles)
7½-minute	1:24,000	2,000 feet	240 meters	7½ × 7½ min.	49 to 70
7½ × 15-minute	1:25,000	about 2,083 feet	250 meters	7½ × 15 min.	98 to 140
Puerto Rico 7½-minute	1:20,000	about 1,667 feet	200 meters	7½ × 7½ min.	71
15-minute	1:62,500	nearly 1 mile	625 meters	15 × 15 min.	197 to 282
Alaska 1:63,360	1:63,360	1 mile	nearly 634 meters	15 × 20 to 36 min.	207 to 281
Intermediate	1:100,000	nearly 1.6 miles	1 kilometer	30 × 60 min.	1568 to 2240
U. S. 1:250,000	1:250,000	nearly 4 miles	2.5 kilometers	1° × 2° or 3°	4,580 to 8,669
U. S. 1:1,000,000	1:1,000,000	nearly 16 miles	10 kilometers	4° × 6°	73,734 to 102,759
Antarctica 1:250,000	1:250,000	nearly 4 miles	2.5 kilometers	1° × 3° to 15°	4,089 to 8,336
Antarctica 1:500,000	1:500,000	nearly 8 miles	5 kilometers	2° × 7½°	28,174 to 30,462

CONTOUR LINES SHOW LAND SHAPES AND ELEVATION

The shape of the land, portrayed by contours, is the distinctive characteristic of topographic maps.

Contours are imaginary lines following the ground surface at a constant elevation above or below sea level.

Contour interval is the elevation difference represented by adjacent contour lines on maps.

Contour intervals depend on ground slope and map scale. Small contour intervals are used for flat areas; larger intervals are used for mountainous terrain.

Supplementary dotted contours, at less than the regular interval, are used in selected flat areas.

Index contours are heavier than others and most have elevation figures.

Relief shading, an overprint giving a three-dimensional impression, is used on selected maps.

Orthophotomaps, which depict terrain and other map features by color-enhanced photographic images, are available for selected areas.

COLORS DISTINGUISH KINDS OF MAP FEATURES

Black is used for manmade or cultural features, such as roads, buildings, names, and boundaries.

Blue is used for water or hydrographic features, such as lakes, rivers, canals, glaciers, and swamps.

Brown is used for relief or hypsographic features—land shapes portrayed by contour lines.

Green is used for woodland cover, with patterns to show scrub, vineyards, or orchards.

Red emphasizes important roads and is used to show public land subdivision lines, land grants, and fence and field lines.

Red tint indicates urban areas, in which only landmark buildings are shown.

Purple is used to show office revision from aerial photographs. The changes are not field checked.

INDEXES SHOW PUBLISHED TOPOGRAPHIC MAPS

Indexes for each State, Puerto Rico and the Virgin Islands of the United States, Guam, American Samoa, and Antarctica show available published maps. Index maps show quadrangle location, name, and survey date. Listed also are special maps and sheets, with prices, map dealers, Federal distribution centers, and map reference libraries, and instructions for ordering maps. Indexes and a booklet describing topographic maps are available free on request.

HOW MAPS CAN BE OBTAINED

Mail orders for maps of areas east of the Mississippi River, including Minnesota, Puerto Rico, the Virgin Islands of the United States, and Antarctica should be addressed to the Branch of Distribution, U. S. Geological Survey, 1200 South Eads Street, Arlington, Virginia 22202. Maps of areas west of the Mississippi River, including Alaska, Hawaii, Louisiana, American Samoa, and Guam should be ordered from the Branch of Distribution, U. S. Geological Survey, Box 25286, Federal Center, Denver, Colorado 80225. A single order combining both eastern and western maps may be placed with either office. Residents of Alaska may order Alaska maps or an index for Alaska from the Distribution Section, U. S. Geological Survey, Federal Building-Box 12, 101 Twelfth Avenue, Fairbanks, Alaska 99701. Order by map name, State, and series. On an order amounting to $300 or more at the list price, a 30-percent discount is allowed. No other discount is applicable. Prepayment is required and must accompany each order. Payment may be made by money order or check payable to the U. S. Geological Survey. Your ZIP code is required.

Sales counters are maintained in the following U. S. Geological Survey offices, where maps of the area may be purchased in person: 1200 South Eads Street, Arlington, Va.; Room 1028, General Services Administration Building, 19th & F Streets NW, Washington, D. C.; 1400 Independence Road, Rolla, Mo.; 345 Middlefield Road, Menlo Park, Calif.; Room 7638, Federal Building, 300 North Los Angeles Street, Los Angeles, Calif.; Room 504, Custom House, 555 Battery Street, San Francisco, Calif.; Building 41, Federal Center, Denver, Colo.; Room 1012, Federal Building, 1961 Stout Street, Denver Colo.; Room 1C45, Federal Building, 1100 Commerce Street, Dallas, Texas; Room 8105, Federal Building, 125 South State Street, Salt Lake City, Utah; Room 1C402, National Center, 12201 Sunrise Valley Drive, Reston, Va.; Room 678, U. S. Court House, West 920 Riverside Avenue, Spokane, Wash.; Room 108, Skyline Building, 508 Second Avenue, Anchorage, Alaska; and Federal Building, 101 Twelfth Avenue, Fairbanks, Alaska.

Commercial dealers sell U. S. Geological Survey maps at their own prices. Names and addresses of dealers are listed in each State index.

INTERIOR—GEOLOGICAL SURVEY, RESTON, VIRGINIA—1978

MILE SCALE 1:24 000

FOOT SCALE 1:62 500

```
25 000 FEET   20 000   15 000   10 000   5000   0   5000
```

Tectonics and Structural Geology

THE EARTH (ZONES AND CHARACTERISTICS)

The Earth's interior is zoned inward on the basis of increasing density: crust, upper and lower mantle, outer and inner cores (Figure 3.1, Table 3.1). The **crust** consists of both **continental** material (granitic; density 2.7–2.8 gm/cc) and **oceanic** material (basaltic or gabbroic; density 3.1 gm/cc), is 7 to 70 kilometers thick, and is thickest under mountain systems. The **mantle**, composed of peridotite with some pyroxene and plagioclase, has a density of 3.3 to 5.6 gm/cc (upper mantle–lower mantle) and is 2,900 km thick. The **outer core** consists of molten iron-nickel metal, has a density of 10 to 12 gm/cc, and is about 2,100 km thick. The **inner core** is composed of solid iron-nickel metal alloy, has a density of 12 to 12.5 gm/cc, and has a radius of 1,250 kilometers. Table 3.1 lists the densities of parts of the Earth and other materials.

The crust and uppermost part of the mantle are collectively called the **lithosphere** and behaves in a brittle fashion toward the surface where faulting occurs, but tends to flow at depth. The **asthenosphere** is below the lithosphere within the upper mantle and behaves plastically (flows) under all natural conditions that we know of, and may contain dispersed magma bodies.

TABLE 3.1   Densities of Parts of the Earth and Related Materials

Material	Parts of the Earth Density (gm/cc)	Natural Materials at Earth Surface Pressure (gm/cc)
Fresh Water, 68° F		0.997
Sea Water, 68° F, 35% salinity		1.025
Ice (Pure)		0.9175
Earth (Ave.)	5.517	Hematite: 5–6
Crust (Ave.)	2.85	
Continental Crust	2.7–2.8	Granite: 2.8 Quartz: 2.65 Orthoclase: 2.55–2.63 Andesine: 2.65–2.68 Calcite: 2.72–2.94

Continued

TABLE 3.1 Continued

Material	Parts of the Earth Density (gm/cc)	Natural Materials at Earth Surface Pressure (gm/cc)
Oceanic Crust	≈3.1	Basalt: 3.0 Labradorite: ≈2.7 Augite: 3.23–3.52 Hornblende: 3.02–3.45
Mantle—Upper	≈3.3	Ultramafic Rocks: ≈3.3–3.4 Anorthite: 2.76 Olivine: 3.92–4.39
Mantle—Lower (possibly similar to stony meteorites)	≈5.6	
Outer Core (possibly similar to nickel-iron meteorites with some silicate materials)	≈10–12	
Inner Core (possibly similar to iron-nickel meteorites with some silicate materials)	≈12–12.5	Iron: 7.875 Nickel: 8.912 Gold: 19.282

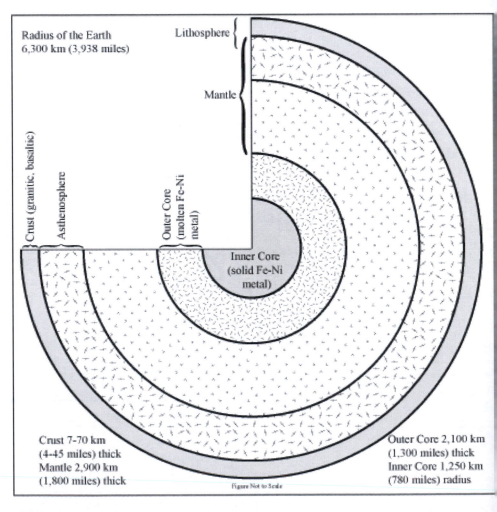

FIGURE 3.1 The Earth's Internal Structure.

CONTINENTAL DRIFT

Alfred Wegener proposed the theory of **continental drift** in the early 1900s. Wegener's theory of continental drift was an effort to explain the matching coastlines of South America and Africa (like pieces of a puzzle). Other evidence supporting the theory of continental drift includes the alignment of structural trends and glacial deposits, the correlation of rock types and ages, and matching fossils, such as the *Glossopteris* (plant) and *Mesosaurus* (reptile), between South America and Africa. Wegener thought the continents of today had to originally have been part of a single supercontinent, which he named Pangaea, at some time in the past. Approximately 180 million years ago, Pangaea began to break apart and separated into two smaller subcontinents. Wegener named the northernmost subcontinent **Laurasia** and the southernmost subcontinent **Gondwanaland**. Laurasia and Gondwanaland, in turn, fragmented into the continents we see today.

Continental drift is still occurring. The documented rates of movement vary from 2 to 17 cm/yr (1–7 in/yr). The North Atlantic Ocean, for instance, is getting wider at a rate of 2 cm/yr. At this rate, the Atlantic Ocean is approximately 10 meters (32 feet) wider than when Columbus crossed the ocean in 1492.

Paleomagnetism, the study of the ancient magnetic orientation of the Earth, also indicates that the continents have drifted with respect to time. As magma or lava with iron-bearing minerals cools and crystallizes, the minerals align to the magnetic north of the time. This orientation will not alter once the magma or lava has solidified. The movement and/or rotation of the continents have disaligned the iron-bearing minerals from the paleomagnetic north direction of ancient rocks relative to the present magnetic north direction, which seems to confirm the theory of continental drift.

Reverse magnetism occurs when the polarity of the Earth's magnetic field changes to the opposite direction (reverses). During times of reverse magnetism, compass needles would point to the south magnetic pole instead of the north magnetic pole. The last magnetic reversal occurred approximately 780,000 years ago (Chernikoff, 1995), when the Earth's polarity returned to its present orientation. Once again, the disalignment of the iron-bearing minerals within rock bodies indicates the continents have drifted with respect to time.

PLATE TECTONICS

Plate tectonics was a theory proposed in the 1960s to explain the actual mechanism of continental drift. The theory states that the Earth's crust is composed of segments of lithosphere that "float" upon the asthenosphere and move in different directions at different rates of speed away from spreading centers. It is thought that the driving force of plate movement is **convection currents** (circulating motions) within the atmosphere and lower mantle. Convection is the upward movement of hotter and usually less dense material, and the adjacent sinking of cooler, denser material. A good visual example of convection can be seen in a pot of boiling oatmeal. The hottest oatmeal rises in the middle of the pot and the cooler oatmeal sinks along the edges.

Plate Boundaries

Plate motion is currently measured with the use of navigation satellites. The edges of plates are recognized as areas of frequent *earthquakes* and *volcanism*. The edge of the Pacific plate is sometimes referred to as the "**Ring of Fire**" because it is surrounded by converging and diverging boundaries that result in numerous episodes of volcanism as well as earthquakes. Tectonic plates collide with one another (converge), pull apart from one another (diverge), or slide horizontally past one another (transform). The plates and plate boundaries of the world are illustrated in Figure 3.2.

Convergent Plate Boundaries

Convergent plate boundaries occur where one or more tectonic plates *collide*. When two plates of differing density, such as continental (2.7–2.8 gm/cc) and oceanic (3.1 gm/cc) plates, collide, the denser oceanic plate is pushed beneath the continental plate in a process called *subduction*. As the leading edge of oceanic plate, with its covering of sediment, is pushed deeper, it first metamorphoses then begins to melt because temperature effectively increases faster than pressure. An approximately *andesitic magma* is created by the partial melting of the subducted plate (as well as some melting and assimilation of continental derived material), rises toward the surface, and usually produces *violent volcanic eruptions*. A *deep trench* forms at the junction of the two plates. *Shallow-, intermediate-,* and *deep-focus earthquakes* occur along the upper surface of the plate being subducted to define what is called the **Benioff zone**, the zone of brittle grinding of the oceanic plate along the edge of the continent-bearing plate. The maximum depth of the earthquake foci effectively defines the regional thickness of the lithosphere. The Benioff zone extends from the trench to depths as great as 700 km (435 miles), and has a slope (dip angle) from less than 30° (faster subduction) to more than 70° (slower subduction toward [underneath]) the continent-bearing plate. The *oceanic plate is destroyed* (metamorphosed and melted) as the plate descends, and the lighter, *continent-bearing plate is deformed by reverse faults and folds*, as a result of *compressional forces*. The type of convergent boundary just discussed was first identified near the west coast of South America, where the Nazca plate is being subducted slowly under the South American plate to form the Andes Mountains, and is often referred to as an Andean Margin.

Mt. Saint Helens, a volcano inactive since 1857, and located along the convergent boundary between the Juan de Fuca plate and the North American plate, explosively erupted on May 18, 1980. The damage caused by this eruption was extensive. Approximately 100 square miles of forest was destroyed and thirty-four people died. Mt. Saint Helens rumbled back to life again in 2004, adding new material to the lava dome already in the crater. In 2005, an eruption of steam and ash occurred. Mt. Saint Helens is located near the site of an earlier volcano, Mt. Mazama. Approximately 6,000 years ago (*Encyclopedia Britannica*, 1986), Mt. Mazama erupted with such force that it was totally destroyed. The explosive and rapid removal of magma caused the volcano to collapse into the magma chamber and form a large **caldera** (a volcanic depression that forms when a magma chamber collapses). The caldera eventually filled with water and is now Crater Lake in Crater Lake National Park, Oregon. Wizard Island is a cinder cone that developed within the caldera after the main eruption.

Earthquakes are also the product of convergent plate boundaries, and some of the most destructive earthquakes in history have occurred at this type of boundary. The top four highest-magnitude earthquakes in recorded history are discussed next.

On December 26, 2004, a "Great Earthquake" (Table 3.2) with a magnitude of 9.0 (sources vary) occurred off the coast of Indonesia in the Indian Ocean basin along the convergent boundary between the Indian and Burma plates. The quaking lasted for several minutes (most quakes last less than a minute) and spawned a massive tsunami that killed over 200,000 people and devastated the coastal areas of many countries. A tsunami is the result of the disturbance of the seafloor by earthquakes or volcanism that radiates outward from the area of disturbance. The waves increase in height and speed as they approach land and cause massive damage.

On Good Friday, 1964, near Anchorage, Alaska, a 9.2-magnitude earthquake also spawned a tsunami. The Pacific plate subducted beneath the North American plate. The ground shook for approximately five minutes. Although the Good Friday earthquake was more powerful than the Indonesian earthquake, the loss of life was much less due to a lower population density in Alaska.

The largest-magnitude earthquake was 9.5 on the Richter scale and occurred on May 22, 1960, in Chile, South America. Many of the dead, approximately 2,000, died as a result of the tsunami generated by the earthquake, not from the earthquake itself. The city of

Valdiva, Chile, was heavily damaged. The tsunami traveled as far away as Hawaii, killing 61 people (Atwater et al., 1999), and Japan, killing 138.

On March 11, 2011, an 8.9-magnitude earthquake struck off the coast of Honshu, Japan, spawning a massive tsunami that killed thousands of people in Japan and one person along the California coast. The epicenter was located approximately 85 miles east of Honshu, Japan, along the convergence of the Pacific plate and the North American plate. The tsunami also caused massive damage to several nuclear power plants along the Japanese coastline, spawning worldwide nuclear concerns.

Divergent Plate Boundaries

Divergent plate boundaries form when two or more plates are pulled apart from one another. The **Mid-Atlantic Ridge** is a much-studied divergent plate boundary. The boundary is formed where the North American and Eurasian plates and the South American and African plates diverge, *along the middle of the Atlantic Ocean.* Analysis of the Mid-Atlantic Ridge has had a great influence on the development of the theory of plate tectonics. The stress involved is *tensional,* which results in **normal** or **vertical faulting** along the rift (divergent) boundary and lower in the lithosphere. Tension reduces pressure and promotes melting, upwelling, and subsequent crystallization of magma that *creates new seafloor.* Many **shallow-focus earthquakes** occur along this type of boundary. **Volcanic ridges** develop symmetrically about the rift and form a **submarine double-mountain chain.** **Iceland** is the only portion of the Mid-Atlantic Ridge that is above sea level. Although most divergent boundaries occur in oceanic crust, some occur in continental crust, such as the **East Africa Rift.** Pangaea was split apart by the development of a series of divergent plate boundaries.

In 1783, Laki, an Icelandic volcano along the Mid-Atlantic rift zone, erupted. The eruption lasted for eight months. The sulfur dioxide gases released by the eruption poisoned the vegetation, which in turn killed the sheep, horses, and cattle the people relied on for food; 10,000 people died from the resulting famine. The sulfur dioxide gases also caused weather changes on a global scale. People as far away as Britain, France, and Germany were also victims of the eruption. Twenty thousand people in Britain alone died by breathing the sulfur dioxide in the atmosphere. Six hundred square kilometers of Iceland were covered by the eruptive lava flows, the largest lava flow in the world during the past 1,000 years (de Castella, 2010).

Eyjafjallajökull, pronounced ay-ya-fyal-la-jo-kult, another Icelandic volcano, erupted in March and April—May 2010, causing a disruption in air travel in Europe for six days.

Transform Plate Boundaries

Transform plate boundaries occur where *plates slide horizontally past one another* along large-scale *strike-slip or transform faults.* The best-known transform plate boundary in the United States is the **San Andreas Fault System** in California. The North American plate (moving southward) and the Pacific plate (moving northward) are sliding past one another. *Shallow-focus earthquakes* that cause a great deal of damage to man-made structures are generated along these boundaries. **Strike-slip or transform faults** develop as a result of *shear stress.* There is generally *little or no volcanism* associated with such boundaries; therefore, *new crust is neither created nor destroyed.* However, because these faults are not straight, there are substantial areas of both compression and extension in adjacent areas.

On April 18, 1906, shortly after 5:00 a.m., an earthquake with a magnitude of 7.9–8.3 (sources vary) struck San Francisco, California. An estimated 3,000 fatalities occurred as a result of the earthquake and resulting fires that razed much of the city to the ground.

On October 17, 1989, an earthquake along the San Andreas Fault Zone occurred in the same region. It measured 7.1 on the Richter Scale of Earthquake Intensity (considered a major earthquake). Sixty-seven people died and a part of Interstate 880 as well as section of the San Francisco Bridge collapsed.

The **Richter Scale of Earthquake Intensity** rates the magnitude (amount of energy released) of earthquakes (Table 3.2) by evaluating the amount of ground motion generated by the quaking. The energy released by earthquakes is increased by a factor of ten as the magnitude increases. Therefore, an earthquake of magnitude 2 is ten times greater than a magnitude 1 earthquake.

The **Modified Mercalli Earthquake Intensity Scale** (Table 3.3) is another method used to rank earthquakes. The Mercalli scale rates earthquakes on the basis of the effects that are readily observable, such as damage to the area affected and the descriptions and sensations of people who lived through an earthquake.

TABLE 3.2   Richter Scale of Earthquake Intensity

Magnitude	Description
2.0 – 2.9	Detected but not felt
3.0 – 3.9	Barely felt
4.0 – 4.9	Minor earthquake
5.0 – 5.9	Slight damage occurs
6.0 – 6.9	Destructive earthquake
7.0 – 7.9	Major earthquake
>8.0	Great earthquake

TABLE 3.3   Mercalli Earthquake Intensity Scale

Intensity	Description
I	Not usually felt, recorded by seismic monitors
II	Felt by few people who aren't moving, liquid sloshes, animals uneasy, chandeliers swing gently
III	Felt by many, feels like a truck is passing by
IV	Noticed indoors, creaking sounds in walls, doors rattle
V	Felt by almost everyone, small amount of damage occurs
VI	Felt by all, damage is slight, cracked plaster, etc.
VII	Some people knocked down, damage to walls greater
VIII	Some difficulty steering vehicles, broken tree limbs, water levels change in wells and springs
IX	Heavy damage, cracks in ground, foundations damaged, subsurface pipes broken
X	Most structures heavily damaged, bridges destroyed, landslides, dams damaged
XI	Railroad tracks warped, underground pipelines broken, buildings collapse
XII	Almost total destruction, objects thrown into air, ground moves in waves

EXERCISE 3.1 PLATE TECTONICS

Complete the following.

1. List any three visibly recognizable <u>physical</u> characteristics that define convergent plate boundaries.

   a. _____

   b. _____

   c. _____

2. List the location of one convergent plate boundary. A convergent plate boundary consists of two different plates. List the names of both plates.

   Plate 1 _____   Plate 2 _____

3. List any three visibly recognizable <u>physical</u> characteristics that define divergent plate boundaries.

   a. _____

   b. _____

   c. _____

4. List the location of one divergent plate boundary. A divergent plate boundary consists of two different plates. List the names of both plates.

   Plate name 1 _____   Plate name 2 _____

5. List one visibly recognizable <u>physical</u> characteristic that defines transform plate boundaries.

   _____

6. What depth of earthquake foci are associated with transform plate boundaries?

   _____

7. What type of stress is associated with transform plate boundaries?

   _____

8. List the location of one transform plate boundary. A transform plate boundary consists of two different plates. List the names of both plates.

   Plate name 1 _____   Plate name 2 _____

9. Assuming a complete lack of knowledge of tectonic plates, where would you describe the location of the Mid-Atlantic Ridge?

10. What part of the Mid-Atlantic Ridge is above sea level? _____

11. What is the "Ring of Fire"?

12. Describe the location of the "Ring of Fire."

13. Look at Figure 3.2, The Tectonic Plates of the World. What **visual** evidence suggests that the continents of South America and Africa were once part of the same land mass?

Alfred Wegener suggested that the continents of today were once part of a supercontinent that broke apart and drifted to their present positions.

14. What was the name of Wegener's supercontinent? _____

15. What are the names of the two smaller plates that Wegener's supercontinent broke into?

   a. _____   b. _____

16. On what four pieces of evidence (not whom) did Wegener base his theory for continental drift, other than the apparent visual fit of South America and Africa?

   a. _____   b. _____

   c. _____   d. _____

17. On Figure 3.2, The Tectonic Plates of the World, label the plates, continents, and major oceans listed below. Label the plate names in black pencil, the continent names in red pencil, and the ocean names in blue pencil. (DO NOT USE PEN.)

Plates (Black)	Continents (Red)	Oceans (Blue)
North American Plate	North America	Pacific Ocean
South American Plate	South America	Atlantic Ocean
African Plate	Africa	Arctic Ocean
Antarctic Plate	Antarctica	Southern or Antarctic Ocean
Nazca Plate	Australia	Indian Ocean
Juan de Fuca Plate	Asia	
Pacific Plate	Europe	
Eurasian Plate		
Indo-Australian Plate		
Caribbean Plate		
Cocos Plate		
Scotia Plate		
Arabian Plate		
Philippine Plate		

18. On Figure 3.2, The Tectonic Plates of the World, outline the different plate boundaries using the following color code.

   Convergent Boundaries—Outline in **Blue Pencil**
   Divergent Boundaries—Outline in **Red Pencil**
   Transform Boundaries—Outline in **Green Pencil**

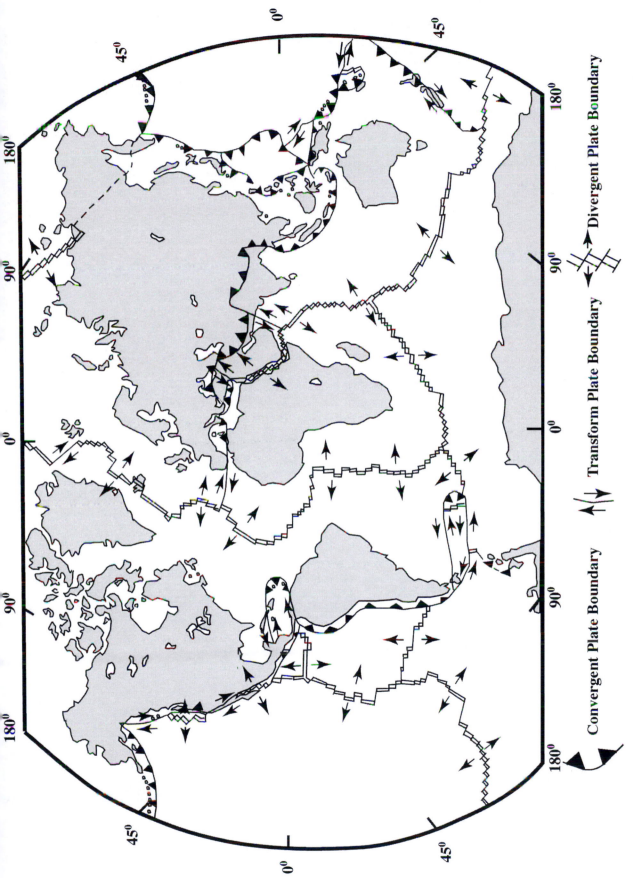

Convergent Plate Boundary

Transform Plate Boundary

Divergent Plate Boundary

FIGURE 3.2 Tectonic Plates of the World, modified from Chernikoff, 1955.

101

2

STRUCTURAL GEOLOGY

Structural geology is the study of the changes in position, or attitude (tilting, bending, stretching, breaking), of bodies of rock achieved by the application of unbalanced (not "hydrostatic"—not equal from all directions) pressures, and varying amounts of heat, after a rock was formed. "Formation" in this sense encompasses the processes of origin of the rock body, such as deposition and lithification (sedimentary rocks), extrusion and solidification (volcanic rocks), magma emplacement and crystallization (plutonic rocks), and even metamorphism (metamorphic rocks). Change in attitude is commonly called **deformation**, and the important point is that the conditions needed for deformation are basically those needed to achieve metamorphism. Deformed rocks are, to a greater or lesser degree, metamorphosed. Some deformed rocks are not noticeably metamorphosed, whereas others are very much changed in texture, mineralogy, and arrangement of their minerals.

Pressure on rocks is of two kinds that act at the same time. **Burial pressure** (lithostatic pressure) is imposed by the weight of the overlying rocks. Burial pressure acts nearly vertically, and is approximately predictable by depth below the Earth's surface. Burial pressure generally builds up and acts slowly and continuously. The second kind of pressure (hydrostatic pressure) experienced by rocks underground is provided by tectonic stresses. At small scales, this kind of pressure can act in any direction, though at large (regional, continental, or global) scale it can be treated as acting nearly horizontally. **Tectonically induced pressures** commonly build up and act rapidly. "Rapid action" in geologic terms means that if we look very closely and measure carefully, we can notice the changes over a few years or tens of years.

If pressure is applied gradually and continuously, and the temperature is high enough, the rock will bend (flow; behave in a plastic or ductile fashion) because the strength of the rock is exceeded. Strength is the tendency of the rock to avoid being deformed, and varies with the temperature and composition of the rock. If pressure is applied rapidly, or temperature is low, or both, and strength is exceeded, the rock will behave in a brittle manner—it will break. Before it breaks, the rock will deform (bend), but much of this deformation disappears after the breakage by means of what is called elastic rebound. **Elastic rebound** means that pieces of the rock snap back into their original form (nearly), so that they can be fitted together to make a body of rock of the original shape. This kind of deformation is said to be "recovered" because the energy used to bend the rock was stored in the rock rather than being used to cause recrystallization and flow (metamorphism). Elastic deformation and the "snap-back" recovery of form produce earthquakes. The energy of an earthquake measured on the Richter scale (Table 3.2) is the energy stored in the rock by elastic deformation. At depths of about 700 km (435 miles), the combination of pressure and temperature appears to make brittle behavior impossible, and rock flows (plastic behavior).

If large situations are considered, it is possible for pressures from all sources and directions to balance each other. This situation is uncommon, at least if time, measured in millions of years, is considered. The idea is useful because it provides a concept of "confining pressure." Any pressure exerted in excess of this balanced or confining pressure will produce deformation of the rocks. The unbalanced part of the pressure usually arises from tectonic stresses, so the study of structural geology is closely related to the study of plate tectonics. The unbalanced pressures can tend to compress the body of rock (wad up the rocks; compression), or extend the rocks (pull the rocks out like taffy; extension). It may seem odd to think of pressure in this way, so maybe it is easier to recognize that pressure is force applied over an area, as in the common expression "pounds [of force] per square inch." The force could be cramming rocks together, or stretching them out (ductile plastic deformation). In a brittle sense, the rocks could break because they are being crushed, or because they are stretched too much.

BRITTLE OR PLASTIC DEFORMATION

The kinds of deformation or geologic structures discussed next are separated into two major groups. Structures produced by brittle deformation are various kinds of **faults** and **joints**. Structures produced by plastic deformation, the result of deep burial, include the folds. These two kinds of deformation can each be further divided into structures produced when unbalanced pressures produce features characteristic of extension and those typical of compression. Separation of the kinds of deformation is useful for discussion and description, but is not entirely realistic. Different kinds of rocks react differently to temperature-pressure conditions that can, in part, be predicted from Bowen's Reaction Series. It is possible and common for one body of rock to have a brittle reaction (break) under a set of unbalanced pressures, while an immediately adjacent rock of a different composition bends (plastic reaction) under the same set of conditions. Furthermore, the set of forces that produces compression in one area will produce extension in adjacent areas.

There is an important point to be considered when thinking about the diagrams used in the definitions and exercises that follow. Geologic structures produced by brittle and plastic (compressional or ductile) deformation affect all kinds of rocks, but are particularly obvious and easy to understand when they affect layered rocks where the major layering has a known original orientation. Volcanic and sedimentary rocks are layered and follow the **Principles of Original Horizontality and Superposition**. Because they are deposited or placed under the influence of gravity, the layers, as originally produced, are nearly horizontal in most instances. Metamorphic rocks are often layered (foliated), but the orientation of the foliation, which is a secondary feature, is usually not horizontal. Diagrams on the following pages illustrate structures imposed on originally horizontal layered rocks (volcanic and/or sedimentary).

STRIKE AND DIP

The terms *strike* and *dip* are used by geologists to describe the orientation of strata. **Strike** is the compass direction of a stratum (bed) that intersects the horizontal. It may help to visualize this if you think of the surface of a lake as a horizontal plane, and a boat ramp as the tilted strata. By looking along the level, horizontal water line on the boat ramp, one could measure its orientation ("direction") with a compass—the strike of the boat ramp. The **dip** of a bed is the amount of tilt, or inclination in degrees, relative to the horizontal. The direction of greatest inclination (steepness) of the boat ramp would occur at right angles to the waterline (strike), so if one stands facing down the boat ramp, one is facing in the **direction of the dip**. This direction is measured with a compass relative to north. The strike, reckoned relative to north, involves only one measurement north (N), east-west (EW), northeast (NE), or northwest (NW). Figure 3.3 illustrates and describes several strike lines drawn at different angles and directions (east and west) of the north line (0°).

There are two measurements for dip: the dip direction, already discussed, and the dip angle. The **dip angle** is the amount of tilt or inclination measured in degrees down from the horizontal, and at right angles (perpendicular) to the strike direction. If strike is N, then the dip direction is either E or W. There can only be one dip direction (it cannot be EW). The dip is recorded with both the dip angle and dip direction. A bed that dips 30 degrees down from the horizontal (0°), toward the northwest direction, is recorded as **30°NW**. An example of the conventional way in North America to record a bed striking 50 degrees east of north (0°) is **N50°E**. The strike can be verbally expressed as being 50 degrees east of north and the dip as 30 degrees to the northwest.

STRIKE AND DIP

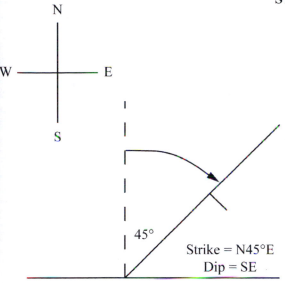

Strike = N45°E
Dip = SE

STRIKE

The strike of a line is measured in degrees towards the East or towards the West of the North direction. The north direction is represented by a vertical line (dashed line on this figure).

The strike is written in a specific format. "N" represents North (the direction measured away from), the angle of the strike in degrees and the compass direction of the angle from north (either towards the east or towards the west.

Therefore, the strike of the figure to the left is **N45°E**.

The strike is 45 degrees east of north.

DIP DIRECTION

The dip direction is determined by a short line that intersects the strike line. The dip line is always drawn perpendicular to the strike line. In the example above, the dip direction is southeast. It is not possible to determine the angle of dip in this figure.

The compass direction of the dip is always perpendicular to the compass direction of the strike.

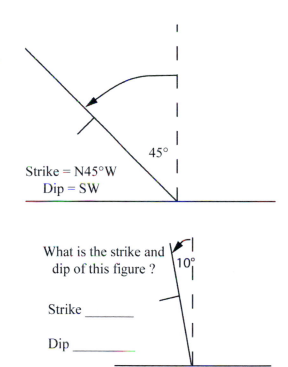

Strike = N45°W
Dip = SW

What is the strike and dip of this figure?

Strike _____

Dip _____

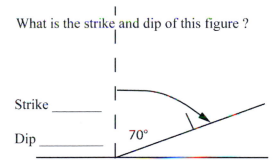

70°

What is the strike and dip of this figure?

10°

Strike _____

Dip _____

When measuring strike, the orientation of the protractor is important. Align the straight edge of the protractor along the north-south line (vertically). Place the point of the protractor at the junction of the north-south and the east-west lines. If the protractor is oriented vertically, the angle can be read as is (where the strike line intersects the protractor). If the protractor is oriented horizontally (the customary orientation) the angle read must be subtracted from 90°.

FIGURE 3.3 Strike and Dip.

EXERCISE 3.2: STRIKE DIRECTION

Strike is the compass direction of a given stratum of rock relative to the horizontal, and relative north. North represents 0°. The horizontal represents a level surface, such as the surface of an imaginary swimming pool, shown in the illustration below. Study the illustration below which shows four boards (that represent strata) that have been positioned vertically in the pool, then answer the following questions. The direction along the top edge of the board is oriented, is the strike direction.

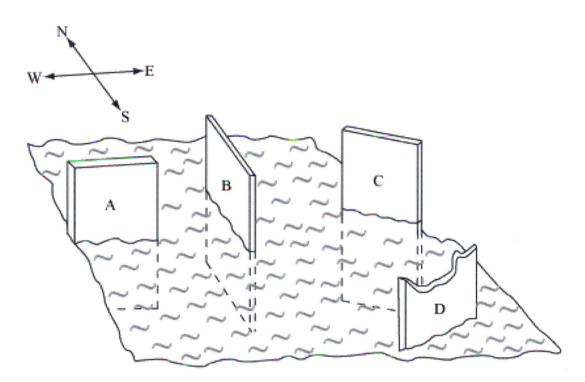

What is the strike direction of board A? _____

What is the strike direction of board B? _____

What is the strike direction of board C? _____

What is the strike direction of board D? _____

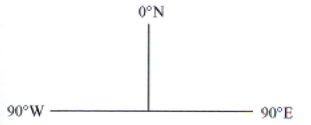

In this course, the strike angle is always measured in degrees from north (0°), up to but not exceeding 90°E or up to but not exceeding 90°W. See figure at left.

EXERCISE 3.3: DIP DIRECTION

Dip Direction is the direction toward which inclined strata or fault planes are dipping (tilted), as visualized by pouring water on a tilted board, and watching which direction the water "runs" downward. The dip direction is always perpendicular (at right angles) to the strike direction.

Dip Angle is the angle of the tilt (dip) of the stratum as measured from the horizontal, such as the surface of our imaginary "swimming pool." All the boards in the drawing below, strike north-south, therefore they will either dip toward the east or toward the west at some angle unless the dip is vertical. Vertical strata are oriented straight up and down and do not have a direction component.

Boards A-D are all striking North-South, therefore the dip direction can only be East or West.
What is the dip of each board?
Remember to use the correct format

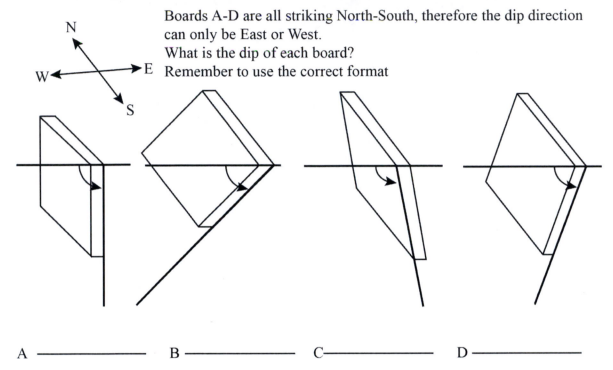

A ——————— B ——————— C ——————— D ———————

What is the dip of lines E-F. Once again assume all strata are striking North-South. Record the dip in the correct format.

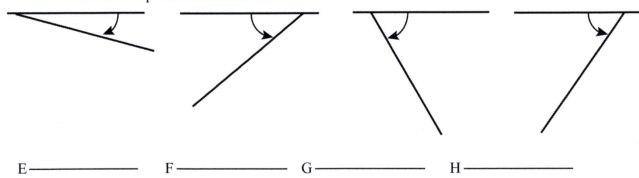

E ——————— F ——————— G ——————— H ———————

EXERCISE 3.4: THE STRIKE AND DIP SYMBOL

The Strike and Dip symbol resembles an elongated letter "T." The strike component is the longer part (top) of the line and the dip component is the shorter part (base) of the line. Study the nine strike or strike and dip symbols and record the strike directions, in degrees, as accurately as possible, and the dip directions. The top of the page is north. Align the protractor vertically with the arrow at the bottom of the strike line. Measure the strike in degrees and record it using the correct format.

Strike _____

Strike _____

Strike _____

Strike _____

Dip _____

Strike _____

Dip _____

Strike _____

Dip _____

Strike _____

Dip _____

Strike _____

Dip _____

Strike _____

Dip _____

BLOCK DIAGRAMS

The structure of subsurface strata can be visualized with the use of three-dimensional representations called **block diagrams** (Figure 3.4). The block diagram shows a **map view** (the surface) and two **cross sections** (the **transverse** or east-west cross section and the **longitudinal** or north-south cross section) that are vertical slices through the Earth, viewed obliquely at the same time. The map view and the age of the strata allow geologists to interpret underground geologic structures. An **outcrop pattern** is a recognizable arrangement (pattern) of strata that is seen if the vegetation and soil are considered to have been removed from the surface. A **contact**, the boundary between two or more strata, is shown on the map and cross-sectional views by a thin black line separating the strata.

GEOLOGIC STRUCTURES

The Earth's crust changes dynamically with time. Rocks originally deposited in horizontal layers can become folded or faulted by compressional, tensional, or shear stress. Examples of structures that are formed by geologic processes include:

1. **Horizontal strata** (Figure 3.5) are (almost always) layers or strata that have not been deformed (or only very gently) since they were deposited (sedimentary) or extruded (volcanic). Horizontal rocks do not have a strike or a dip component (i.e., an absence of structure). Only a single stratum, the youngest, is seen on the map view. Cross-sectional views show horizontal and parallel contacts between the beds.

2. **Inclined (tilted) strata** (Figure 3.5) are layers of strata that are tilted from the horizontal, up to but not exceeding 90 degrees. The strike of the tilted strata is parallel to the contacts (stripes) seen on the map view. The "stripes," non-repeating (all of different ages), comprise the outcrop pattern. The strata always dip toward the youngest bed unless the rocks have been overturned. None of the structural situations in this manual concern overturned beds.

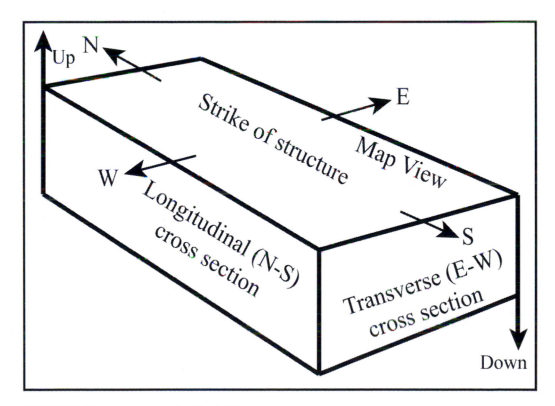

FIGURE 3.4 Anatomy of a Block Diagram.

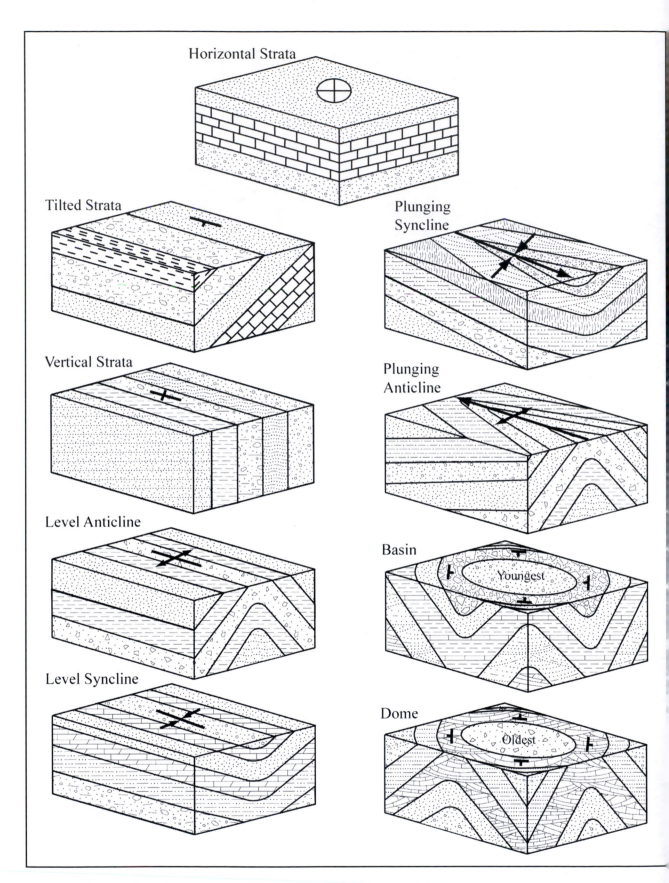

FIGURE 3.5 Geologic Structures.

3. **Vertical strata** (Figure 3.5) are layers of strata that dip 90 degrees from the horizontal (standing on edge). Vertical strata have a strike that is parallel to the contacts between the layers, and a dip of 90 degrees, but no dip direction. Parallel non-repeating stripes (outcrop pattern) are seen on the map view.

4. **Nonplunging folds (anticlinal and synclinal)** are layers of strata that are bent or crumpled. Folds result when compressional forces act on portions of the Earth's crust. The compression causes the crust to arch, bend upward, or bend downward (form troughs). The resulting arches or troughs are called folds. An upward arch is an **anticline** and a trough is a **syncline**.

   The axis of a fold is the line at which the dip angle changes direction. Folds have axial planes, "invisible" planes that bisect the fold. Symmetrical folds have vertical axial planes, and thus the same degree of dip on both sides of the axis. Each side of the fold is a mirror image of the other. An asymmetrical fold has a tilted axial plane, and thus has different dip angles on either side of the axis. Each side is NOT a mirror image of the other, so that the width of the outcrop "stripes" seen on the map view are different on either side of the axis. The strike of individual layers within anticlines and synclines parallel the stripes on the map view. The stripes are only seen after erosion has removed the uppermost part of the anticline and syncline structures (Figure 3.5).

   a. **Level anticlines** (Figure 3.5) are arch-shaped folds and the limbs dip away from the axis. When the anticlines are eroded, the map view shows parallel repeating stripes. The axis is located in the centermost, "oldest" stripe. The transverse cross section shows arched strata and the longitudinal cross section shows level, parallel, horizontal beds.

   b. **Level synclines** (Figure 3.5) are downwardly bent strata (troughs). The limbs dip inward toward the axis. When synclines are eroded, their map views also show an outcrop pattern of parallel, repeating stripes, but the axis runs through the "youngest" (centermost) stripe. The transverse cross section shows a trough and the longitudinal cross section shows level, parallel, horizontal beds.

5. **Plunging folds (anticlinal and synclinal)** are folds that are themselves inclined or tilted; in other words, their axial planes are tilted and plunge into the Earth. Converging and diverging symmetrical repetition of strata in the map view indicate the presence of plunging folds. Chevrons (V- or S-shaped) patterns are seen on the map view as given strata wind back and forth on the surface. Anticlines and synclines alternate, and plunging folds are the most common.

   a. **Plunging anticlines** (Figure 3.5). The direction of convergence of the stripes, seen on the map view, will point in the direction of the plunge of the fold (toward the nose, the pointy end of the chevron). The transverse cross section shows the strata arched and the longitudinal cross section shows parallel, tilted beds. The beds tilt in the same direction as the plunge.

   b. **Plunging Synclines** (Figure 3.5). The direction of divergence of the stripes, seen on the map view, will point in the direction of the plunge of the fold (away from the nose, the wider part of the chevron). The transverse cross section shows the strata downwardly bent and the longitudinal cross section shows parallel, tilted beds. The beds tilt in the same direction as the plunge.

6. Doubly plunging folds, **domes** or quaquaversal anticlines, and **basins** (Figure 3.5), are arched-shaped and downwardly bent structures, respectively. If the strata form "rings" or "bevels" in the map view, either a dome or a basin is present underground. Domes and basins are called doubly plunging folds because all cross sections look like anticlines or synclines.

   a. **Domes** (Figure 3.5) are identified by concentric, irregularly shaped ellipses or rings on the map view, with the oldest ring in the center of the structure. All cross sections are anticlinal, arch-shaped.

   b. **Basins** (Figure 3.5) are identified by concentric, irregularly shaped ellipses or rings on the map view, with the youngest ring in the center of the structure. All cross sections are synclinal, downwardly folded.

MAP SYMBOLS

The identification of structures, man-made and natural (roads, strata), on maps is accomplished with the use of standardized symbols. The map symbols used in this section to indicate the geologic structures of the Earth's crust are illustrated in Figure 3.6.

FAULTS

A **fault** is a fracture (evidence of brittle failure) in the Earth's crust along which measureable relative movement has occurred. Earthquakes are the result of the sudden release of stress elastically stored in a rock body, and usually indicate movement (slippage) along fault lines. The subsurface point where the movement that generated the earthquake occurred is called the focus and the point directly above the focus, on the Earth's surface, is called the epicenter.

Faults can have either *vertical* or *inclined fault planes* (plane of slippage). For faults with inclined fault planes, the **hanging wall** of a fault is the side that would hang over your head if you could walk within or along the fault zone. The **footwall** of the fault is the side that you would walk upon within or along the fault zone. The side of the fault that moves upward is the **upthrown** block and the side that moves downward is the **downthrown** block. Figure 3.7 illustrates the four different faults we will discuss:

1. **Vertical faults** (Figure 3.7) are generally caused by tectonic extensional stresses, at depth. Although uncommon, they may develop in near-surface situations, as adjustments to deep-seated tectonic stress. The fault plane is vertical and neither hanging walls nor footwalls develop.

2. **Normal faults** (Figure 3.7) are the result of tensional stress at or near the Earth's surface, where pressures are too low to allow rocks to stretch without breaking, and they

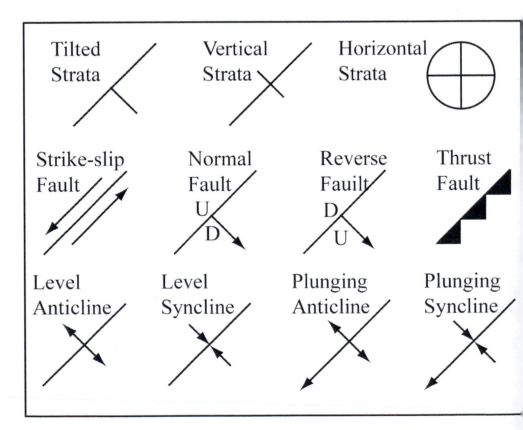

FIGURE 3.6 Geologic Map Symbols.

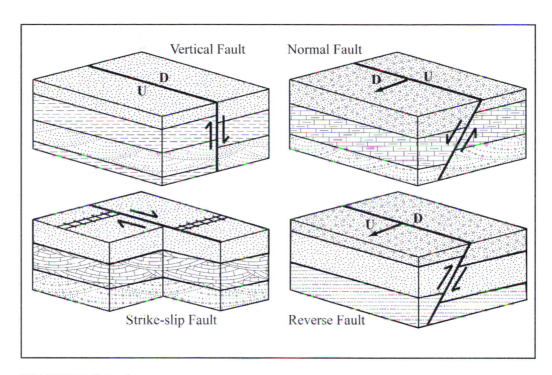

FIGURE 3.7 Faults.

have inclined fault planes. The rock breaks and gravity causes the hanging wall to move down the dip of the fault plane or zone relative to the footwall (dip-slip motion). Normal faults are associated with divergent plate boundaries (spreading centers), rifts, and areas of thinning of the Earth's crust, such as in the Great Basin region.

3. **Reverse faults** (Figure 3.7) are the result of compressional stress. The hanging wall is forced upward relative to the footwall. Reverse faults are most commonly found at convergent plate boundaries. A **thrust fault** is a low-angle reverse fault.

4. **Strike-slip (transform) faults** (Figure 3.7) are the result of horizontal movement along a plane and are the result of shear stress. The strike-slip fault is so named because the motion is parallel to the strike of the fault. Fault planes may be inclined but are more commonly near vertical, and there is no cross-sectional offsetting of strata because the opposite sides of the fault slide horizontally past one another. Figure 3.7 illustrates a right lateral strike-slip fault. The same strata or feature, in this case a railroad track, on the opposite side of the fault has moved to right when you are facing the fault. The opposite is true for a left lateral strike-slip fault. Strike-slip faults are located at the junction of transform plate boundaries such as the San Andreas Fault System, and along zones within plates where unequal motions of different parts of plates are accommodated (e.g., Trans-Pecos, West Texas).

CONTACTS

There are two types of contacts between strata. A **conformable contact** is a boundary between rock units (strata) where there is no record of lost time through erosion or lack of deposition. An **unconformable contact** is a boundary between rock units where time has gone unrecorded due either to erosion or nondeposition; the rock units below the boundary are much older than those above the boundary. These contacts, *when buried*, are called **unconformities**. The symbol for an unconformity on a cross section is a wavy line (Figure 3.8). There is no map symbol for an unconformity.

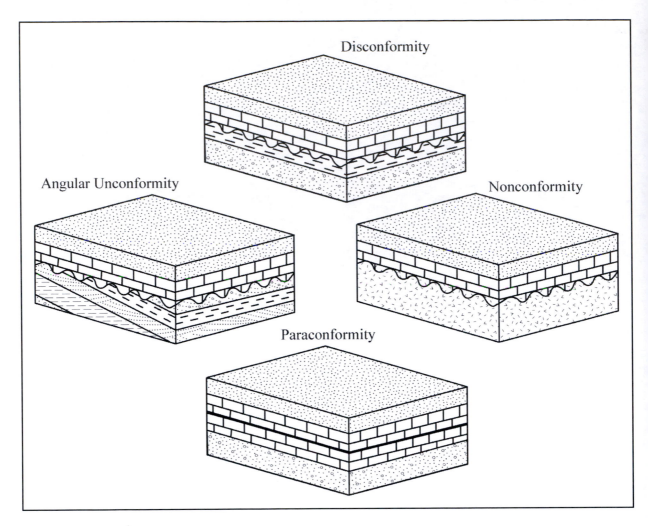

FIGURE 3.8 Unconrormities.

UNCONFORMITIES

1. **Disconformity** (Figure 3.8) refers to situations where the strata above and below a surface of erosion belong to the same rock family (sedimentary) and are at the same attitude (parallel) to one another. The unconformable surface shows relief due to the erosion.

2. **Nonconformity** (Figure 3.8) refers to situations where the strata above the surface of erosion are layered sedimentary or volcanic rock and those below are much older plutonic igneous rock and/or metamorphic rock.

3. **Angular unconformity** (Figure 3.8) refers to situations where the strata above and below a surface of erosion meet at an angle; the older strata below are tilted, bent, or faulted and the younger strata above the unconformable surface are horizontal (unless the rocks have been secondarily overturned).

4. **Paraconformity** (Figure 3.8) refers to situations where rock units of different ages are separated by a surface of *nondeposition*. The strata above and below the unconformable surface are at the same attitude (parallel to one another). Other evidence (usually fossils that establish age) must be used to determine that time has passed between the deposition of the layers because no obvious erosion occurred between the episodes of deposition.

> Geologic structures can be found on the Earth and Space Science website (http://ess.lamar.edu). Click on the People tab, Staff, Karen M. Woods, Teaching, Physical Geology Lab, Geologic Structures.

EXERCISE 3.5: FAULTS

Label the following on the figure below:

On the map view
U - on the upthrown block   D - on the downthrown block

On the Cross-section
Draw arrows on opposite sides of the fault indicating direction of movement

A.
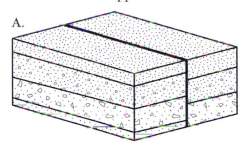

What type of fault is figure "A"?

What type of fault is figure "B"?

What type of fault is figure "C"?

Label the following on each of the figures below:

On the map view   What type of fault is figure "D"?
U - on the upthrown block   D - on the downthrown block
Draw an arrow indicating the tilt direction of the fault plane

On the Cross-section
H - on the hanging wall   F - on the foot wall
Draw arrows on opposite sides of the fault indicating direction of movement

B.
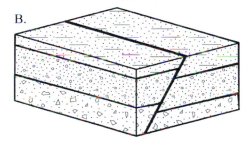

C.

Draw arrows on opposite sides of the fault indicating
direction of movement on the figure below.

What stress results in normal faults?

What plate boundary are normal faults associated with?

D.
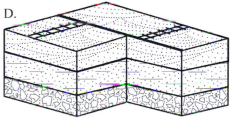

What stress results in reverse faults?

What plate boundary are reverse faults associated with?

What stress results in strike-slip faults?

What stress results in vertical faults?

What plate boundary are strike-slip
faults associated with?

What plate boundary are vertical faults associated with?

EXERCISE 3.6: UNCONFORMITIES

Define unconformity _____

_____

_____

Define disconformity _____

_____

_____

Define nonconformity _____

_____

_____

Define angular unconformity _____

_____

_____

Define paraconformity _____

_____

_____

What symbolizes an unconformity on a cross-section? _____

Draw a disconformity in the box above

Draw a paraconformity in the box above

Draw a nonconformity in the box above

Draw an angular unconformity in the box above

EXERCISE 3.7: OPTIONAL

Complete the map views on the following block diagrams.
Place the **correct map symbol**, in the correct place **on the map view**.
On the lines provided, **identify (label) each structure**.

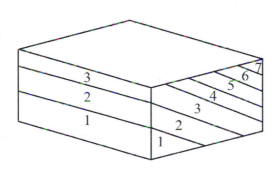

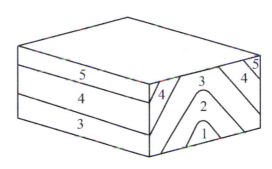

_____

_____

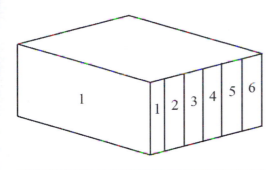

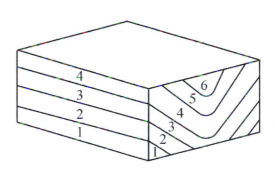

_____

_____

Complete the cross-sections on the following block diagrams.
On the lines provided, **identify (label) each structure.**

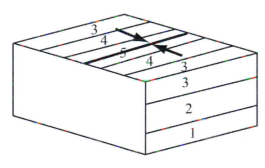

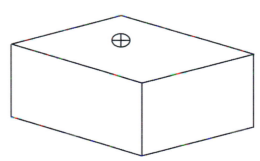

_____

_____

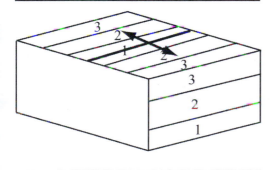

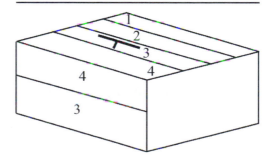

_____

_____

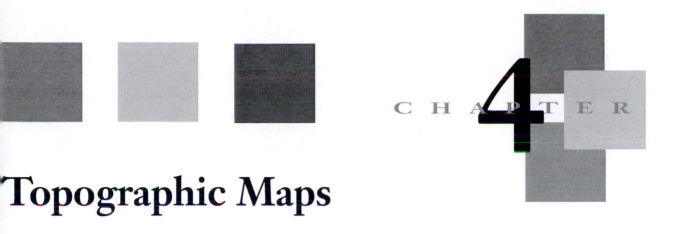

Topographic Maps

INTRODUCTION

Everyone is familiar with road maps, which show the location of roads, cities, major rivers, and other features. A **topographic map** is a two-dimensional (flat) representation of a three-dimensional surface that shows the location of every natural and man-made object or feature within the map area. Topographic maps are not only useful to geologists. Hunters, hikers, engineers, contractors, foresters, architects, and so on also use topographic maps because of their accuracy. Hills, valleys, ridges, depressions, other features, and different man-made structures such as roads, buildings, dams, and so forth are depicted by the use of special symbols. A key to the symbols on topographic maps is located at the end of the colored section of this manual.

ELEVATION

Elevation is the height of the land above mean sea level (0 feet) (depths below sea level are labeled with a negative sign in front of the number (e.g., -200) to indicate the depth below sea level). The arithmetic difference in elevation between any two elevations on a topographic map is called **relief**. Relief can be expressed as **total/maximum relief**, the difference in elevation between the highest and lowest elevation on the map, or as **local relief**, the difference in elevation between a hilltop and its adjacent valley.

The United States Department of the Interior, Coasts and Geodetic Survey, National Oceanographic and Atmospheric Administration (NOAA), the United States Geological Survey (USGS), and other agencies and organizations are instrumental in the very accurate measurement of specific locations in the United States. Permanent brass markers are set in the ground at various locations. The plates, called **bench marks**, are engraved with the exact elevation or a reference code for that location. Bench marks are indicated on maps by the letters BM, followed by a number indicating the elevation at that point. Elevations that have been surveyed into a given place in the map area but are not marked by permanent markers are called **spot elevations** and are indicated on maps by the letter "x" followed by the elevation (e.g., x250).

CONTOURS

Contours are lines drawn on topographic maps that represent equal elevations above (or below) a datum point, usually near sea level (0 feet). Contours represent three-dimensional features such as hills, valleys, ridges, and so forth as lines on paper.

121

The **contour interval** on a topographic map (once determined) is the vertical difference in elevation between any two adjacent contour lines. Once a contour interval is determined for a specific map, it will remain constant throughout. For example, contour lines on a map with a contour interval of 10 (feet or meters) would have values that are multiples of 10 (0, 10, 20, 30, etc.). The contour interval selected for a particular area is dependent on how flat or how hilly (mountainous) the area is. The more mountainous the terrain, the larger the contour interval, and vice versa. **Index contour lines** are labeled with the elevation and are drawn thicker on topographic maps. In general, every fifth contour line on a map is an index contour line. If the contour interval were 10 feet (or meters), the index contour interval would be 50 feet (multiply the contour interval [10] by 5).

INTERPRETATION OF CONTOUR LINES

Depressions, hills, valleys, slopes, ridges, and so forth are represented on topographic maps by various contour line patterns (Figure 4.1). Closed contour lines, irregularly shaped rings, represent either **hills** (Figure 4.1a) or **depressions** (Figure 4.1b). The contour rings for hills and the contour rings for depressions are easily distinguishable on a topographic map. The contour rings representing **depressions** are hachured (include tick marks) on the inside of the rings and the smallest ring (centermost) represents the lowest contoured elevation, but not necessarily the lowest point of the depression.

The contour rings representing **hills** lack hachure marks and the smallest centermost ring represents the highest contoured elevation of the hill, although not necessarily the highest point on the map.

Slopes (Figure 4.1c) are represented by contour lines that are more or less parallel to one another.

Both **ridges** and **valleys** are represented by contour lines that bend back and forth. Contour lines that point toward the lower elevations represent ridges (Figure 4.1d) (high ground). Contour lines that point toward the higher elevations represent **valleys** (Figure 4.1d), the

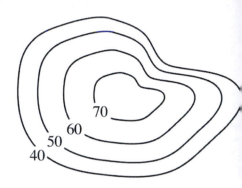

FIGURE 4.1a Hill.

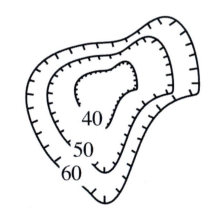

FIGURE 4.1b Depression.

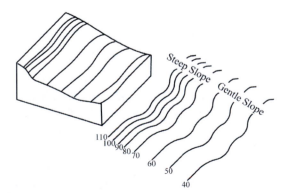

FIGURE 4.1c Slope.

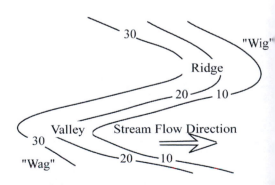

FIGURE 4.1d Valley and Ridge.

low-lying land between hills or mountain ranges that usually has a stream flowing through it. Streams are either **intermittant** (flow part of the time) or **perennial** (flow all the time). Valley shapes are features of topography that help to interpret the history of landscapes (Figure 4.2). A V-shaped valley is typical of a stream in the youthful stage of the cycle of erosion. U-shaped valleys are the result of erosion by glaciers, and squared-off valleys are in the mature stage of the cycle of erosion. The closer the lines are drawn to one another, the steeper the slope of the land. Conversely, the farther apart the contour lines are drawn to one another, the gentler the slope.

If the contour "Wag" terminates as a pointed V, the valley has a **V-shaped transverse profile**, typical of streams in the **youthful** stage of the cycle of erosion

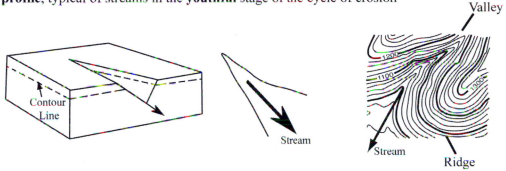

If the contour "Wag" terminus is rounded, the valley has a **U-shaped transverse profile**, and a **glacial valley** is indicated.

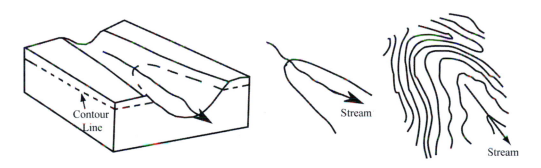

If the contour "Wag" is squared, the valley has a **flat bottom, a flood plain**, and the valley is in the **mature** stage of the cycle of erosion.

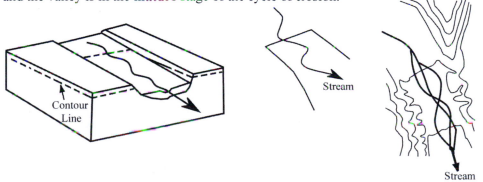

FIGURE 4.2 Valley Shapes.

DRAWING CONTOURS

The purpose of drawing contour lines is to produce a two-dimensional representation of the topography (ups and downs) of the landscape of a given area. We will create contour maps utilizing a traditional method. When surveying instruments were purely optical, surveyors would establish spot elevations as accurately as possible. The elevations and coordinate positions of a network of high and low points would be plotted as a base map. Another member of the team, usually an artist, would draw in the contour lines around the surveyed spot elevations relative to a given contour interval. This manual will provide spot elevations. Think of yourself as the artist.

One problem of drawing contours is that the spot elevations are chosen because of their aerial distribution, and because they provide information about maximum, minimum, and average elevations in the area of the proposed map. You, as the "artist" in lab, cannot actually see the landscape topography. The first step is to consider the contour interval (CI), the difference in the elevation of two adjacent contour lines. The particular contour interval chosen depends on the amount of relief and the amount of detail required.

Look at Figure 4.3. What is the total relief of the upper map? What is the total relief of the lower map? What contour interval would you choose? Why?

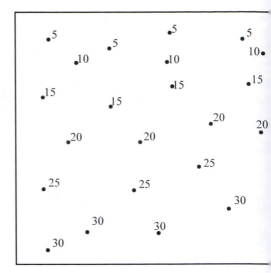

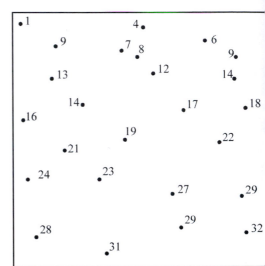

FIGURE 4.3 Beginning Contours.

To accurately place contour lines, the "artist" must use his or her best interpretation (best guess) of the location the contour line should be drawn at. This is called **interpolation**. For example, if the contour interval is 10 feet and you are given the spot elevations a 11 foot and 9 foot, you would draw the 10-foot contour line midway between them. Ar comes into play later when the map is almost done and when the student has gained in sight into the nature of the lay of the land. Of course the actual mapmakers draw contou lines while in the field; thus they see the hills, valleys, and other features. You, in the lab will be at a disadvantage. Be imaginative. Go back to Figure 4.3 and contour the maps. The contour interval for both maps is 5 feet.

CONTOURING RULES

When drawing contour lines, some basic rules must be followed:

1. Contour lines connect points of equal elevation, as defined by the contour interval.
2. A contour line will never split or divide.
3. Contour lines never just end. They will continue and join up with a contour line o an adjacent map (if the contour interval is compatible).

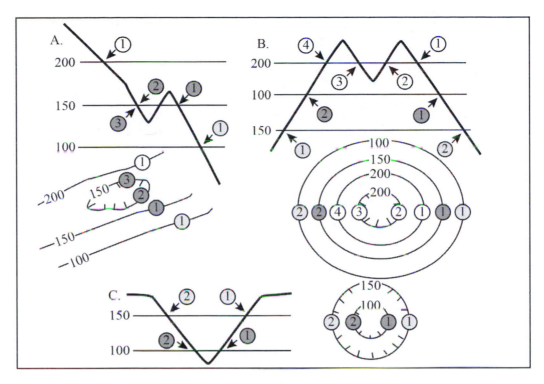

FIGURE 4.4 Repeating Contour Lines A. On the slope of a hill **B.** On a volcanic peak. **C.** A depression on level ground. 1–4 represent the number of times a contour line will repeat in the direction the arrows indicate.

4. One contour line will never intersect another contour line. (There are exceptions, such as in the case of an overhanging cliff, but they will not occur in this lab manual).

5. Contour lines will bend and point uphill (toward higher elevations) when crossing a valley. The bend will either be V-shaped (youthful), squared (mature), or U-shaped (glacial) at the head of the valley.

6. Contour lines representing ridges (high ground between valleys) will bend and point downhill (toward lower elevations).

7. Contour lines that are widely spaced represent a gentle slope.

8. Contour lines that are closely spaced represent a steep slope.

9. Hills or mountain peaks are represented by irregularly shaped concentric circles (contour rings).

10. Depressions are represented by irregularly shaped concentric circles (contour rings) with hachure (tick) marks drawn on the inside of the rings.

11. Contour lines can repeat in certain circumstances (e.g., a depression on a hill, or a volcanic peak with a crater). Figure 4.4 illustrates these situations.

TOPOGRAPHIC PROFILE

A **topographic profile** (Figure 4.5) is a cross-sectional representation of the lay of the land as if the hills and valleys are seen from a distance. To draw a topographic profile, the "artist" must first contour a map, then select a desired cross section, such as A-A' (Figure 4.5). After a cross section has been chosen, the elevation of each contour line that crosses the A-A' is transferred to the vertical scale of the cross section determined by the maximum and minimum elevations that cross the A-A' line. To transfer the points, take a piece of blank paper and line it up with the cross-section line on the contour map. Make marks on the blank

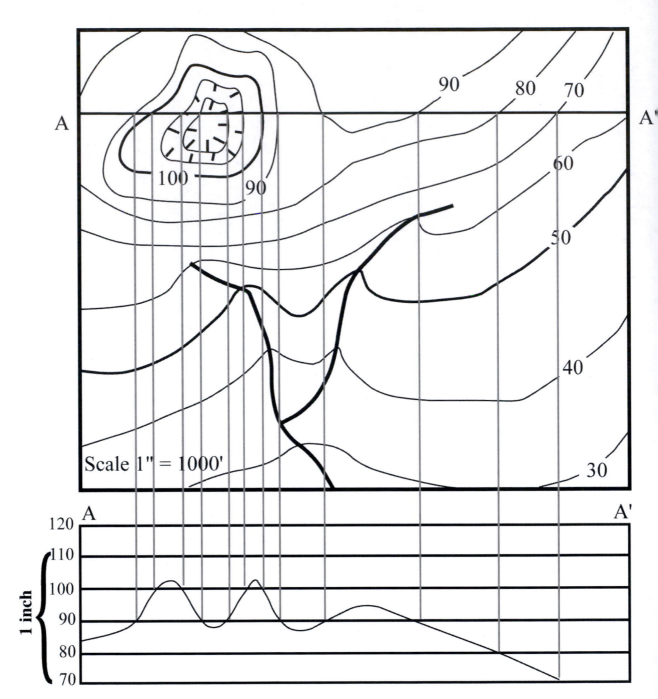

FIGURE 4.5 Topographic profile. The topographic profile (A-A') illustrates repetition of contours at hills and depressions. This is a profile of a vocano. Note the contour lines "V" (point) in the uphill direction as they cross the rivers.

paper template where the contours cross the A-A' line and at each end of the section line. Record the representative elevations at the marks. Now move the paper template to the cross section. Using the same spacing, make a dot on the cross section in accordance with the elevations on the vertical scale. Once all the points are marked, connect the dots with a "smooth" (not jagged) line. Be on the "lookout" for typical topographic features such a hills, valleys, depressions, and so on. Draw a hill with an arch shape to indicate that it is hill. A distinct notch can be used to indicate the position of a stream valley.

VERTICAL EXAGGERATION

The vertical exaggeration of a topographic profile is the amount the profile has been "stretched" vertically relative to the horizontal A-A' distance. It is necessary to exaggerate the vertical scale in order to emphasize the shape of the hills and valleys. **Vertical exaggeration** is a ratio of the horizontal scale of the map (in feet/inch) divided by the amount of feet in one measured inch on the vertical scale. For example, if the horizontal (map) scale is 500 ft/in and one measured inch on the vertical scale equals 250 feet (250 ft/in), the units (feet/inch) will cancel and the vertical exaggeration will be 2x (2 times), or 500 ft/in divided by 250 ft/in. Vertical exaggeration must be greater than one. It is essential for both the map scale and the vertical scale to have the same units.

Look at Figure 4.5. The horizontal (map) scale is one inch equals 1,000 feet. The relief within one measured inch on the vertical scale is 40 feet (110 ft − 70 ft). Therefore, the vertical exaggeration is 25x (1,000 ft/in divided by 40 ft/in).

MAP SCALES

All maps have one or more scales that are used to determine distances from point to point. The **ratio scale** is a ratio of a given measure on the map to the number of equivalent measures on the ground that covers the same distance. The ratio 1:62500 means that every inch on the map equals 62,000 inches on the ground (a little less than a mile). A ratio scale of 1:24000 means that every measured inch on the map equals 24,000 inches on the ground.

The ratio scale (inches to inches) is usually converted to feet per inch, a more familiar scale. Therefore, the ratio 1:24000 converts to 1 inch is equal to 2,000 feet (24,000 inches divided by 12 inches per foot = 2,000 ft/in) on the ground.

Another type of map scale is the **graphic** or **bar scale**. It is illustrated by graduated bars in feet, miles, and kilometers. Look at the bar scale labeled miles on the 7.5-minute series topographic map provided. The total scale represents 2 miles. The part of the bar scale to the left of the zero (0) is subdivided into 10 increments, each of which represents a tenth of a mile. Similarly, the feet scale is subdivided into five 200-foot intervals to the left of the zero (0) and thousand-foot intervals to the right of the zero (0).

A third type of map scale is called the **verbal scale**. The verbal scale is simply an expression of the scale in words. For example, the verbal conversion of 1:24000 would be expressed as "one inch on the map equals twenty-four thousand inches on the ground, or about a third of a mile," and 1:62500 is expressed as "one inch on the map is equal to approximately one mile on the ground."

EXERCISE 4.1: BEGINNING CONTOURS

Answer the questions for the corresponding figures.

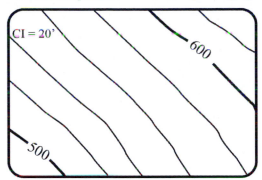

What topographic structure is shown above?

What are the labeled and thicker contour lines called?

What are the thinner contour lines called?

What is the index contour interval?

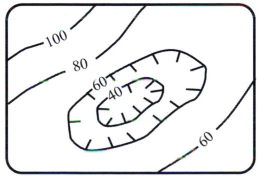

What topographic structure is shown above?

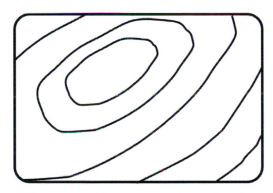

What topographic structures are shown above?

How deep is the depression (rim to bottom)??

What is the contour interval?

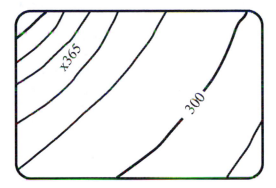

What topographic structure is shown above?
Be Specific.

What is the contour interval?

What did you use to determine the contour interval?

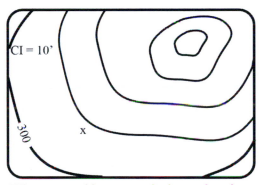

What topographic structure is shown above?

What is the elevation of the structure?

What is the elevation of the " x"?

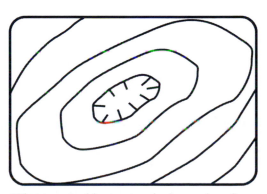

What topographic structure is shown above?
Be imaginative.

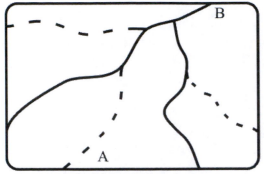

The figure above illustrates stream.

What kind of stream is illustrated by solid lines?

What kind of stream is illustrated by dashed lines?

Which letter indicates the higher ground?

What compass direction are the rivers flowing?

Label all the contour lines with thier elevations on the figure below

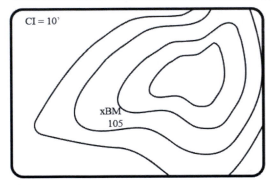

What topograpic structure is illustrated above?

What is the elevation of the highest point of the structure (extimated)?

What is the total relief of this area?

Label the Index Contour lines in the spaces provided.

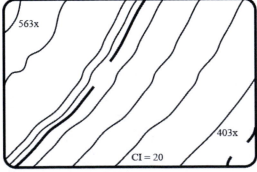

What is the total relief of this structure?

What topograpic structure is illustrated above?

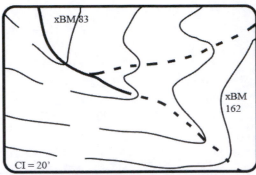

Label the contour lines in the spaces provided.
Notice the bench marks. Remember, contour lines "V" (point uphill) when crossing a river

Label the contour lines in the spaces provided on the

figure below.

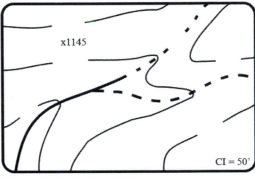

What is the total relief of this area?

What topographis structures are illustrated above?

(Not the streams)

What compass direction are the streams flowing?

Label the contour lines in the spaces provided.

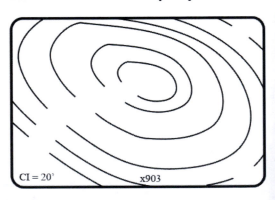

What is the elevation of the highest point of the structure?

What topographic structure is illustrated above?

EXERCISE 4.2: MORE CONTOURING

Shown at right is a map of a peninsula of land for which spot elevations have been determined. Notice sea level is at an elevation of 0 feet. Using a CI of 10 feet, see if you can construct topographic map of the peninsula.

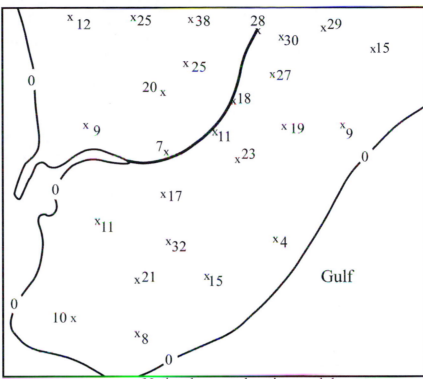

Using a CI of 20', see if you can create the appropriate topographic map expressed by these elevations. Label the contour lines.

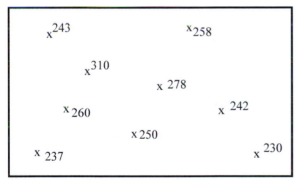

Notice the spot elevations and the streams on the map below. Using a CI of 10', see if you can create the appropriate topographic map expressed by these elevations. Label the lines.

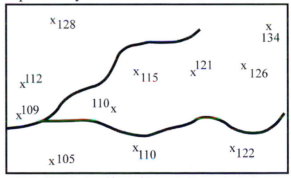

Notice the spot elevations on the map below. Using a CI of 10', see if you can create the appropriate topographic map expressed by these elevations. Label the contour lines.

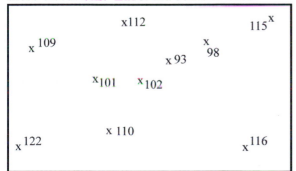

Notice the spot elevations on the map below. Using a CI of 10', see if you can create the appropriate topographic map expressed by these elevations. Label the contour lines.

EXERCISE 4.3: TOPOGRAPHIC PROFILE

Doug Point Mountain

900

1000

800

1100

900

800

Lamar Lake

A —— A'

1200

x1379

1300

Lake Olivia

800

900

800

Kay Lake

800

800

1" = 2000'

Observe the contour lines shown. How many different lines (elevations) must you consider for this map?

On the vertical scale below.

1. Label the elevations in the blanks provided.

2. Draw the topographic profile A-A', as accurately as you can.

Remember, the surface of the lakes are horizontal, ridges arc upward and valley arc downward.

Elevation

_____   _____

_____   _____

_____   _____

_____   _____

_____   _____

_____   _____

_____   _____

Calculate the vertical exaggeration. Show your work.

EXERCISE 4.4: CONTOURING & TOPOGRAPHIC PROFILE

1. Contour the map using the spot elevations provided.
2. Construct the Topographic Profile A-A'
3. Calculate the Vertical Exaggeration. Show your work.

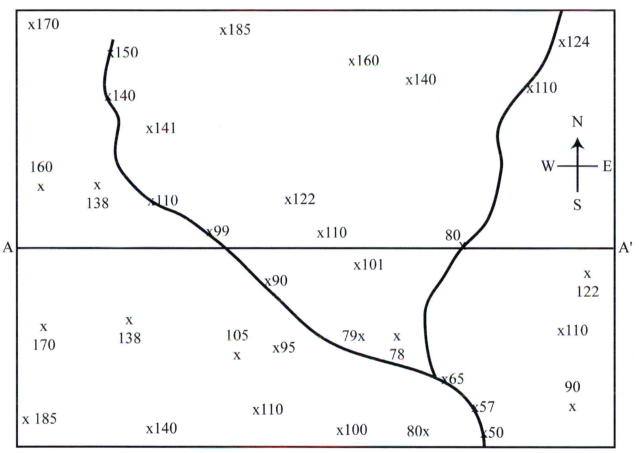

Scale 1"=320'
Contour Interval 20 feet

COORDINATE SYSTEMS AND MAP LOCATIONS

Topographic maps are sometimes referred to as quadrangle maps. A **quadrangle map** (four angles, sides) is a portion of the Earth's surface that is bounded by longitudinal lines on the eastern and western edges of a map and by latitudinal lines on the northern and southern edges of a map. Latitudes and longitudes are coordinates. Topographic maps use the intersection of coordinates to indicate the location of the map area on the Earth's surface. Although other coordinate systems exist, the latitude/longitude coordinate system will be taught in this course.

Latitude consists of unequal-length east-west lines (circling the Earth), expressed in degrees north or south of the equator (0°), up to but not exceeding 90°N or 90°S (Figure 4.6). **Longitude** consists of equal-length north-south lines that converge at the poles and expand at the

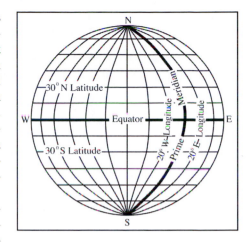

FIGURE 4.6 Lattitude and Longitude Corrdinate System.

equator (Figure 4.6). Longitudes are expressed in degrees east or west of the prime meridian (0°), up to but not exceeding 180°E or 180°W. The **prime meridian** is the longitude line established by international treaty in 1884 that passes from pole to pole through Greenwich, England, and is also referred to as the **Greenwich meridian** (Figure 4.6). The prime meridian is the basis of international time zones.

Both latitude and longitude lines are expressed as degrees (arcs) measured from the center of the Earth. Latitudes are measured in arcs north or south of the equator and longitudes are measured east or west of the prime meridian. A degree of an arc is composed of sixty units referred to as minutes (60') and a minute is composed of sixty units referred to as seconds (60"), neither of which is associated with time. The length in degrees, minutes, and seconds of one side of a topographic map is called the **map series** (7.5-minute series, 15-minute series, etc.) and indicates how much latitude and longitude is covered on an individual topographic map. A representation of some important characteristics of topographic maps is illustrated in Figure 4.7.

OTHER COORDINATE SYSTEMS

Highway (road) maps use a coordinate system that combines letters with numbers to aid in the location of a given city. Other coordinate systems include the **State Plane Coordinate System**, the **Universal Transverse Mercator System (UTM)**, and the **Township, Section and Range System**. Some maps, not usually quadrangle maps, are produced using **state plane coordinates**. These coordinates are determined by the given states. If a state is large enough, more than one set of state plane coordinates may be used to subdivide the state. New Mexico, for example, has three state plane coordinate areas, Northern, Central, and Southern. The Universal Transverse Mercator (UTM) coordinate system is useful to navigators because it is conformal—coordinates intersect at right angles—and thus it provides true N, S, E, and W compass directions. Conformal maps show the true shape of geographical features. Longitude and latitude are also expressed as true N-S and E-W straight lines. Quadrangle maps have UTM values, as well as latitudes and longitudes, printed along the map edges. The third coordinate system is the Township, Section and Range System (TSR). This system is the "flat Earth" coordinate and land-survey system, which is to say it would work very well if the Earth were flat. Because the Earth is not flat, TSR coordinates established in different regions do not meet very closely, if at all.

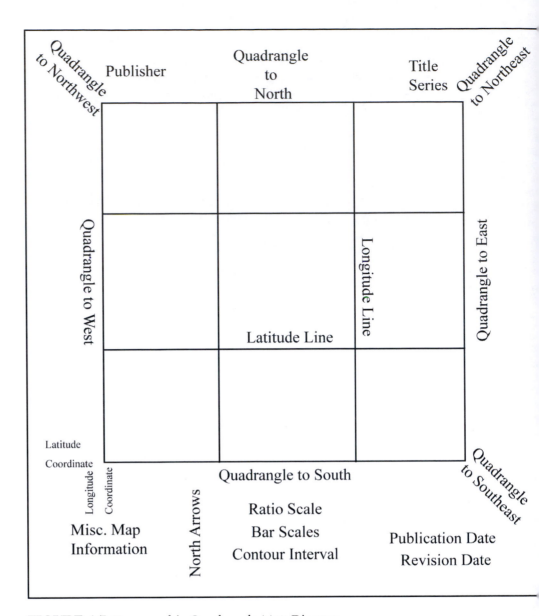

FIGURE 4.7 Topographic Quadrangle Map Diagram.

MAP DIRECTIONS

Maps are oriented in relation to true north. There are actually three different "north" directions designated on topographic maps. **True** or **geographic north** is where the Earth rotates around its axis. **Grid north** always points in a northerly direction, parallel to the longitude lines. **Magnetic north** is the direction a compass needle points to, a magnetic anomaly in the Earth itself, and not true north. The **magnetic declination** of a map is the angle formed between the direction of true north and magnetic north, and it changes as one moves west (or east). Magnetic declination has both an angular and directional component. The direction is determined by the location of the mapped area. If the mapped area is west of the magnetic north line, then the magnetic declination has an east direction; if the mapped area is east of the magnetic north line, then the magnetic declination has a west direction.

SIMPLE CONVERSIONS

Maps are produced either using the metric system (centimeters, meters, kilometers) or the English system (inches, feet, miles). Conversions from one to another are commonly necessary when using topographic maps. The following is a short list of commonly used measurements and conversions.

12 inches = 1 foot	5,280 feet = 1 mile	0.6 miles = 1 kilometer
1.6 kilometers = 1 mile	1 degree = 60 minutes	1 minute = 60 seconds
1:24,000 = 2,000 feet	1:62,500 = 5,208.333 feet	

CALCULATING LATITUDE AND LONGITUDE COORDINATES

Geologists utilize maps to specifically locate points via latitude and longitude. To find these coordinates, the geologist must know the length in inches between two adjacent latitude coordinates or two longitude coordinates as well as the difference in degrees, minutes, and seconds between the two chosen coordinates. The Kingston, Rhode Island, 7.5-Minute Series map will be used for coordinate determinations.

Latitude of Gravel Pit

In the southwest corner of the map is a gravel pit, just above Cedar Island. The chosen point is the symbol for a quarry or open pit mine (a pair of crossed pick-axes; see Topographic Map Symbols Key) next to the letter "G" in the word *gravel*. We will determine both the latitude and longitude coordinate of the quarry.

The first step in determining latitude coordinates is to calculate the number of seconds represented by every measured inch as latitude changes (as you move up and down the map).

1. Choose any two adjacent latitude coordinates and take the difference.

    Example: 41°22′30″N–41°21′30″N = 1 minute (60 seconds).

2. Measure the distance between the two coordinates in inches.

    3 inches

3. Divide the number of seconds by the length in inches.

    60 seconds/3 inches = 20 sec/inch

    So, every vertically measured inch is equal to 20 seconds of a degree.

4. Measure the shortest distance to the point chosen—either up from the bottom or down from the top.

    In this example the shortest distance is from the bottom edge of the map (Latitude 41°21′30″N).

    4.65 inches

5. Multiply the distance by the number of seconds per inch and convert back into minutes and seconds format.

    20 sec/in × 4.65 in = 93 seconds = 1′33″

6. The coordinates increase in the northward direction; therefore the previous answer must be added to the coordinate originally measured from—in this example, from the bottom edge.

    41°21′30″ + 1′33″ = 41°23′03″N

Longitude of Gravel Pit

The first step in determining longitude coordinates is to calculate the number of seconds represented by every measured inch as longitude changes (as you move right to left across the map).

1. Choose any two adjacent longitude coordinates and take the difference.

Example: 71° 37′30″ W—71° 35′00″N = 2′30″ (150 seconds).

2. Measure the distance between the two coordinates in inches.

5.7 inches

3. Divide the number of seconds by the length in inches.

150 seconds/5.7 inches = 26.3 sec/inch

So, every horizontally measured inch is equal to 26.3 seconds of a degree.

4. Measure the shortest distance to the point chosen—either from the right edge or from the left edge.

In this example the shortest distance is from the left edge of the map (Longitude 71° 37′30″ W).

1.55 inches

5. Multiply the distance by the number of seconds per inch and convert back into minutes and seconds format.

26.3 sec/in × 1.55 in = 40.7 seconds = 41″

6. The coordinates decrease in the eastward direction; therefore the previous answer must be subtracted from the coordinate originally measured from—in this example from the left (western) edge.

71° 37′30″ − 41″ = 71° 36′49″ W

The coordinates of the gravel pit are 41° 21′03″ N and 71° 36′49″ W.

EXERCISE 4.5: TOPOGRAPHIC MAP READING

Answer the following questions for the map provided.

What is the name of this map?

Where is the name of the map located?

What is the latitude coordinate along the northern boundary of the map?

What is the latitude coordinate along the southern boundary of the map?

What is the longitude coordinate along the western boundary of the map?

What is the longitude coordinate along the eastern boundary of the map?

How many minutes of a degree are represented between the two latitude coordinates on the northern and southern boundaries?

How many degrees are represented between the two latitude coordinates on the northern and southern boundaries?

How many minutes of a degree are represented between the two longitude coordinates on the western and eastern boundaries?

How many degrees are represented between the two longitude coordinates on the western and eastern boundaries?

Is this map a 7.5-minute, 15-minute, or 30-minute series?

When was this map first published?

Was this map revised? If yes, when?

What is the ratio scale of this map?

How many miles/inch are there on this map?

How many kilometers/inch are there on this map?

What is the contour interval of this map?

What is the name of the quadrangle map north of this one?

What is the name of the quadrangle map south of this one?

What is the name of the quadrangle map east of this one?

What is the name of the quadrangle map northwest of this one?

KINGSTON, RHODE ISLAND, 7.5-MINUTE SERIES MAP

Answer the following questions for the map provided.

What is the elevation of the hill south of the "K" in the word "Kingstown" (near the center of the map)?

What is the approximate elevation of the church (northeast corner of the map) directly below Kingston Road?

What is the elevation of Worden Pond?

What is the local relief between the top of Great Neck and Worden Pond?

In what compass direction does Whitehorn Brook flow (near Larkin Pond, upper center)?

What is the magnetic declination of this map?

EXERCISE 4.6: BEAUMONT EAST, TEXAS, 7.5-MINUTE SERIES

Answer the following questions for the map provided.

What is the name of this map?

What is the latitude coordinate along the northern boundary of the map?

What is the latitude coordinate along the southern boundary of the map?

What is the longitude coordinate along the western boundary of the map?

What is the longitude coordinate along the eastern boundary of the map?

When was this map first published?

Was this map revised? If yes, when?

What is the ratio scale of this map?

What is the contour interval of this map?

What is the calculated index contour interval of this map?

What is the approximate latitude of the Plummer Building (round building with flag) on the Lamar University Campus?

What is the approximate longitude of the Plummer Building (round building with flag) on the Lamar University Campus?

What is the approximate latitude of the sand pit south of Rose City?

What is the approximate longitude of the sand pit south of Rose City?

What is the approximate elevation of Spindletop School (south of Lamar)?

What is the highest elevation near Vidor?

What is the name of the quadrangle map north of this one?

What is the name of the quadrangle map west of this one?

What is the magnetic declination of this map?

What is the distance in kilometers between the Plummer Building and St. Anthony's School? (Show your work.)

What is the distance in miles between the Plummer Building and St. Anthony's School? (Show your work.)

What is the distance in feet between the BMx22 (bottom, middle of map) and the screen at the Old Drive-In Theater near Twin Lakes?

Draw the map symbols for the following:
Small cemetery   Church   Mine tunnel   Well (other than water)

Look at the bar scales of this map. Why is the 0 (zero) not placed at the end of the bars?

Rivers, Streams, and Landscapes

WATER

Water (H_2O) is one of Earth's most important resources. Every living organism depends on water for survival. Ninety-seven percent of the Earth's water contains significant amounts of dissolved salts. Fresh water, the remaining 3 percent, occurs in the form of ice caps, glaciers, groundwater (2 percent) and streams, lakes, and swamps (1 percent).

Water is circulated throughout the Earth's atmosphere and hydrosphere via the **water (hydrologic) cycle**. Surface evaporation from oceans, seas, lakes, rivers, and soils; the transpiration of plants; and the perspiration and waste of animals introduce water into the air as water vapor. As cool air rises, it becomes saturated: water vapor condenses into clouds, and is released as precipitation that falls back to the Earth and continues the cycle. The United States Geological Survey has developed a "Water Science for Schools" website that may be of interest: http://ga.water.usgs.gov/edu/index.html. This website also has links to other related sites.

The amount of water vapor air can hold is dependent on the temperature of the air (the warmer the air, the more water vapor it can hold). The percentage of the amount of water vapor in a parcel of air relative to the amount it can hold at a given temperature represents the **relative humidity** of that air. A cold surface will "sweat" (except in very dry climates) as the surface cools the immediately adjacent air to a temperature below the **dew point**, the temperature at which condensation begins. In general, when air cools below the dew point, dew forms on the surfaces on the ground and as precipitates on condensation nuclei (microsurfaces) in the atmosphere. This can happen near ground (fog), but more commonly it occurs as updraughts of warmer air rise and cool, by physical expansion (adiabatic expansion). Clouds, masses of water droplets (or tiny ice crystals), form in this way. If and when the droplets or crystals become large enough, they fall to the Earth in the form of precipitation (rain, snow, hail, etc.), taking with them the dust on which the droplets formed. Most precipitation on land dampens soil, then evaporates back into the air. Some precipitation infiltrates into the ground and becomes groundwater, and some of it flows across the ground and into streams, as runoff, which then flows toward lower elevations and ultimately to the oceans and seas.

STREAM AND RIVERS: GENERAL TERMINOLOGY

Fluvial (stream) features are described by a variety of terms. The beginning of a river or stream is called the **headwaters**. A river flows from its headwaters downhill to its **mouth**, the lowest part of a river. The flowing water of river erodes the landscape to form a **valley**.

Tributaries are side streams that join the main river, and distributaries (associated with rivers near or at base level) are streams that diverge from the main river. The total combined area drained by a river and all its tributaries is called the drainage basin or watershed. The high ground that separates one river system from another or one drainage basin from another is called a divide. The North American Continental Divide extends irregularly through the Rocky Mountains from Alaska through Mexico, and separates the United States into two main drainage basins, each of which is separated into several smaller drainage basins. Precipitation west of the Rocky Mountains has potential to flow into the Pacific Ocean, whereas precipitation that falls east of the Rockies has potential to flow into the Gulf of Mexico or the Atlantic Ocean. Streams can be perennial (permanent), those that flow all the time, or intermittent, those that only flow part of the time. Perennial streams are illustrated on maps with solid blue lines, and intermittent streams are illustrated by dashed and dotted lines on maps (see Topographic Map Symbols, end of color section). Streams that end abruptly on a map, in reality, sink into the ground and are called disappearing streams. Most disappearing streams are associated with karst topography (see next chapter). The gradient of a river is the difference in elevation between the headwaters and the mouth divided by the length in miles. A river that loses half a foot per mile has a low gradient, whereas a river that loses 50 feet per mile has a very high gradient. Old streams generally have low gradients and young streams generally have high gradients.

RIVER MORPHOLOGY AND CHARACTERISTICS

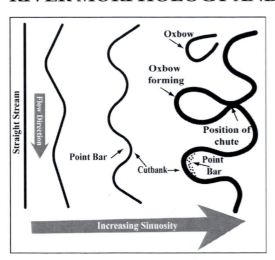

FIGURE 5.1 Sinuosity Straight (youthful stage) stream, to highly meandering (old stage) stream, cutbank, point bar, chute, and an immature oxbow.

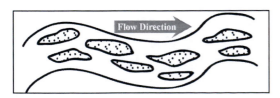

FIGURE 5.2 Braided River.

In general, there are three river morphologies: straight (1) meandering (2), or braided (3). Straight streams (1) usually have a high gradient and little to no flood plain. When a river loses gradient because it has encountered a temporary or absolute base level, the river will meander (swing back and forth). The amount of meandering is discussed in terms of sinuosity (Figure 5.1), which can be expressed as the length of the river along its channel (where the water flows when the river is not in flood) compared to the distance covered by the river as the "crow flies" (the straight-line distance from the headwaters to the mouth). A highly sinuous meandering river (2) has bends that almost meet themselves "coming and going." Meandering rivers may erode through the narrow neck of land between one meander and another, in times of flood to create a chute and cut off and abandon the old meander as flooding subsides.

The abandoned meander becomes an oxbow, and, if filled with water, it is referred to as an *oxbow lake* on the flood plain. Over time, an oxbow lake may turn into an *oxbow marsh*, as the water content decreases, and is referred to as a *wooded oxbow* when dry. Sediments are deposited on the inside of meanders creating features called point bars, mainly on the downstream part of the inside of the bend, and it is in these deposits that fossils may be preserved. Erosion of older sediments takes place on the outside of the bends, creating cutbanks.

Most rivers have only one active channel at a time. Braided streams (3) (Figure 5.2) are overburdened with sediment and have many crisscrossing channels active at the same time.

FLOOD PLAINS AND FLUVIAL DEPOSITS

When a river is in flood and flows out of its channel, it flows across a vegetated flood plain. The water flowing above the vegetation flows more slowly than the water in the deeper channel. The decrease in velocity results in deposition, particularly close to the river's edge. The resulting thicker deposits that are immediately parallel to the channel are **natural levees** (Figure 5.3). Natural levees allow a meandering river to build up the central part of the flood plain, until the river is flowing at an elevation that is higher than its own flood plain. Man-made levees are commonly built along rivers that flood frequently in hopes of containing them in their channels. The development of natural levees can hinder tributaries from joining the main river. This often results in the tributary flowing parallel to the main river along the outside of the natural levee un-

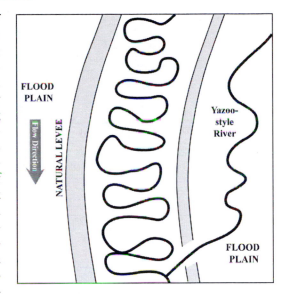

FIGURE 5.3 **Stream Features** Flood plain, natural levee, and yazoo tributary.

til it can breach the levee and unite with the main river. The parallel stream is called a **yazoo tributary** (Figure 5.3), named for the Yazoo River in west-central Mississippi. The Yazoo River is a tributary of the Mississippi for about 175 miles before finally joining it.

The comparatively flat area parallel to the river where the river stores sediment between floods is the **flood plain**. Therefore, flood plains (Figure 5.3) are called that for the simple reason that when the river overflows its banks, it may top or break through natural levees and flood large areas of low land between the levees and the valley walls. For most of the length of a river, the flood plain can be regarded as a temporary storage reservoir for sediment in transit, even if it may take hundreds of years for the river to pick it up again. Nevertheless, **fluvial** (river deposited) sediments are sometimes preserved, particularly those deposited near a coast, and are an important source for many kinds of plant and animal fossils.

Flowing water has two properties that are important to the erosion and deposition of sediment, capacity and competence. **Capacity** is the total amount of sediment a river can carry at any given time. **Competence** is the maximum size of grains or clasts that a river can carry at a given time. The velocity (speed of flow) is the most important factor determining both capacity and competence. When the gradient of a river decreases rapidly, both the capacity and competence decrease and there is deposition. The first sediments deposited are the coarser sediments (gravel or sand), which then grade into the finer sand and silt-sized sediments. The water has to be nearly or completely quiet (unmoving) to deposit clay-sized particles. Natural levees are composed of coarser sediments than those deposited further from the river on the flood plain. However, the most striking examples of deposition occur when a river comes out of mountainous terrain into flatlands, to deposit an alluvial fan. **Alluvial fans** are easily recognized on topographic maps. The contour lines tend to develop a scalloped shape. When a river enters a standing body of water such as a lake or an ocean, the sediments form a **delta**. These bodies of sediment tend to be triangular in longitudinal cross sections and fan-shaped cones in map views, but alluvial fans have more steeply sloping surfaces than deltas (2–5 degrees for alluvial fans, compared to a fraction of a degree for most deltas), and much coarser sediments.

RIVERS AND EROSION: DEVELOPMENT OF LANDSCAPES

Most precipitation that falls to Earth simply evaporates and returns to the atmosphere. Some runs off across the landscapes to form rivers and streams. What happens next is largely a matter of the slope or gradient of the land surface over which the water flows. On a level surface, the water spreads out equally in all directions and would simply soak underground. If the land has a gentle slope, the water moves across it downhill as a sheet. Sheet flow can be observed in a heavy downpour if the local topography is very smooth and simple. However, water usually flows into erosional irregularities, first into *rivulets*, then *rivulets into streams*, and then *streams into rivers*. If the gradient is sufficient enough, the flowing water will have enough velocity to create a new gully. Erosion is enhanced if there is not much vegetation to slow the water by friction, and if soil is friable (loose). Sand and silt are sediment sizes that fall within the competence range of many streams, and are easy to move.

Most gullies begin forming by simple down-cutting. If erosion is very rapid, or if the top of the gully walls are protected by vegetation, or are armored by sediments too coarse or too clayey for new streams to erode, the down-cutting will continue for some time. Gorges are small canyons with near-vertical sides, and a box canyon (rincon) is a canyon with a closed upper end. Usually the walls of the gully have rivulets running down into it and in cross section the gully has a V-shape. V-shaped valleys are characteristic of streams and rivers that are down-cutting fairly rapidly, but, at the same time, they gradually get wider. If the stream encounters alternatively hard and softer layers as it cuts down, the "V" will be stair-stepped. This is called differential erosion. The harder layers (difficult to erode) make breaks in the slopes—the slope of the valley side is more gentle above the harder layers, steeper where the layer has been cut through, and gentler again below the hard layer.

When downward erosion slows, either because the river has encountered an especially hard layer it cannot erode, has met a temporary base level, or because the stream no longer has enough capacity to do more than move sediment supplied by the tributaries, the stream will start to build a flood plain. The original V-shaped valley develops a flat bottom. Contour line "V's" point in the upstream direction as they cross a stream in a V-shaped valley, but as a flood plain develops, the contour lines that cross a flood plain are squared.

The erosive power of rivers can be immense. Streams eventually cut through all kinds of solid rock. The Colorado River cut the mile-deep Grand Canyon, in the last 3 million years, through limestone, sandstone, shale, granite, gneiss, and schist. As the Kaibab Plateau slowly uplifted and faulted, the Colorado River, initially fault-controlled, cut deeper and deeper into the relatively horizontal underlying rock. The erosional process of the river formed cliffs (breaks in the slopes for harder layers) and slopes along the (steep) valley wall as it cut downward. Sandstone and limestone, relatively hard rocks compared to shale, form cliffs in the Grand Canyon walls. The shale beds form gentler slopes between the harder beds of sediment. The scouring action of slow-flowing valley glaciers will erode a **U-shaped** valley because the ice can scrape high along the valley walls. One can tell how thick (deep) the glacier was by the height where the rounding stops on the side of the valley.

FLOODING ALONG THE MISSISSIPPI RIVER SYSTEM

In 1927, the Mississippi River breached the existing levee system in 145 places and flooded 26,000 square miles (Kosar, 2005). The floodwaters reached up to 30 feet deep in places. Ten states, in all, were affected by the flooding, resulting in $400 million in damages and the deaths of 246 people. The severity and scope of the 1927 flood resulted in the building of a larger levee system as well as several floodways and spillways as part of the Flood Control Act of 1928. The Birds Point-New Madrid Floodway (north of Cairo, Illinois), the Morganza Floodway (near Morganza, Louisiana), and the Bonnet Carré Spillway (25 miles north of New Orleans) were all constructed to divert water from the Mississippi and reduce the flood stages (http://www.mvn.usace.army.mil/pao/bro/misstrib.htm). **Floodways** are areas between a river's levee and a setback levee (a levee built back from the river) reserved for diverting floodwaters to protect populated areas. **Spillways** are passages where excess water can be diverted from the main river.

The "Flood of 1993" along the Mississippi and Missouri rivers and their tributaries was also devastating. An unusually high rate of precipitation in the Midwest and man-made levees along the riverbanks contributed to the magnitude of the flood. The residents of the threatened areas attempted to prevent the spread of water over their lands and into their homes as the water levels rose, but many homes, farms, and businesses were destroyed when the levees breached. Approximately fifty people died, and damage was estimated to be more than $10 billion (Mairson, 1994).

In April and May 2011, flooding along the Mississippi River and its tributaries resulted in billions of dollars in damages to homes, businesses, and farms. On May 3, 2011, the Army Corps of Engineers blasted 2 miles of the levee at the Birds Point-New Madrid Floodway, flooding 130,000 acres of farmland, to protect Cairo, Illinois, and to prevent the rest of the levee system from breaching. On May 15, for the first time in almost four decades, the Morganza Floodway (20 miles long and 5 miles wide; http://www.mvn.usace.army.mil/bcarre/morganza.asp) was opened to divert part of the Mississippi into the Atchafalaya Basin and lessen the flooding in Baton Rouge and New Orleans. The Bonnet Carré was also opened, diverting water into Lake Pontchartrain.

STREAM STAGES ("AGES") OF DEVELOPMENT

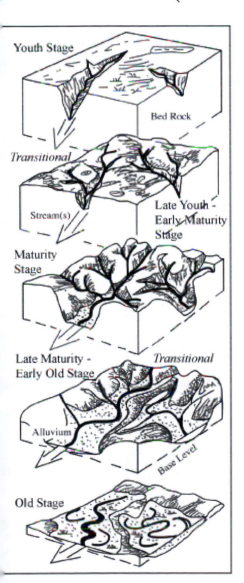

FIGURE 5.4 Stream Stages.

The development of streams and rivers has long been discussed in terms of "ages": youth, maturity, and old age. The conditions described are real, but there is little relationship between each stage and actual time, because a complex of variables makes no two terrains or river systems alike. The "ages" of stream and river development are better discussed as stages in development. Characteristic features of the landscape and the river itself define a given stage. The characteristic features are evident on topographic maps.

Youth Stage: Rivers in the **youth stage** (Figure 5.4) are generally *small*; have *steep gradients*; are *down-cutting, widening* their valleys and eroding headward; have *frequent waterfalls* and *few tributaries*; are rather *straight*; and are *confined in V-shaped valleys*. The *uplands (divides) are broad and poorly drained*, with *frequent lakes* and *swamps*.

Late Youth-Early Maturity (Transitional) Stage: Rivers can be in a stage that is transitional between youth and maturity (Figure 5.4). Such rivers have *increased length* due to headward erosion, *gentler gradients, more tributaries, sinuous channels (beginnings of meandering)*, and have developed a more complicated drainage basin, but the *valley remains relatively straight* and V-shaped. The *uplands are narrower* and better drained when in youth but *lakes* and *marshes still exist*.

Maturity Stage: Rivers in the **maturity stage** (Figure 5.4) have their *maximum number of tributaries*, and hence the most complicated drainage basin. The *river is closer to base level* than in the youth stage and thus begins to *meander* (becomes more sinuous). The stream erodes *cutbanks* on the outside of each meander, *deposits point bars* on the inside of each meander, and *builds a narrow flood plain* as the meanders migrate as a wave downstream. There are no waterfalls but the river may have rapids produced by irregularities in the bottom of the stream that produces turbulent flow. The *uplands are all in slope*, and the *divides between one drainage basin and another tend to be ridges* (sometimes called hogbacks). The *land has its maximum relief* and *no natural lakes* or *swamps* occur. All the land is drained.

Late Maturity-Early Old (Transitional) Stage: Rivers can be in a stage that is transitional between maturity and old (Figure 5.4). Such rivers have *flat, terraced areas between divides*, and the streams have *broad, flat flood plains*.

Old Stage: Rivers in the **old stage** (Figure 5.4) have their *maximum length* and the *fewest number of tributaries*. They *meander extensively* because they flow across an essentially level flood plain. *Oxbow lakes* form when the river cuts off meanders during floods. The *land is relatively flat*, at or near base level, and is *poorly drained*. Oxbows and swamps are common and the *divides that separate one river from another are low*.

Stream Rejuvenation

Rejuvenated streams are streams or rivers that have been forced back one or more stages—literally, "made youthful again." Rejuvenated streams are the result of an increase in the erosive power of the stream, possibly caused by uplift of the areas over which they flow, by eustatic (worldwide) sea level fall, or by an increase of capacity (from increased rainfall that increases runoff due to climate change). An entrenched (deeply dug in) river is the result of slow rejuvenation in which uplift of the land occurred at the same rate as downward erosion.

LANDSCAPE STAGES ("AGES") OF DEVELOPMENT

Topography, the ups and downs of the land, is determined by the same factors that control the stages of development of stream and river erosion: amount of time the land has been exposed to the ravages of climate, composition of the materials being eroded, and regional tectonic activities. Landscape features are identifiable in the field and are recognizable on topographic maps.

Youthful landscapes (Figure 5.5) are characterized by the presence of a few streams contained in separate and distinct valleys, and by broad, perhaps somewhat flat-topped upland (divides) between the streams. The rivers are in the youthful stage of development.

Mature landscapes (Figure 5.5) are characterized by eroded divides. All the land is in slope and the rivers begin to meander and develop flood plains (youth to early maturity stage).

Late mature landscapes are characterized by the presence of wide, flat lowlands (flood plains) that develop along rivers. The rivers are in the mature stage of development, and are highly sinuous (meander widely) with well-developed flood plains.

Old landscapes (Figure 5.5) are characterized by the virtual erosion (elimination) of the divides between river systems. The lowlands are nearly at the same elevation as the rivers and the rivers are generally old and well established.

STREAM DRAINAGE PATTERNS

When rivers are viewed from the air they make distinctive patterns on the Earth's surface called drainage patterns. When there is no structural control (no underground "opinion"), the aerial pattern produced by rivulets joining creeks, creeks joining streams, and streams joining rivers will resemble the branching of trees, and is called a **dendritic drainage pattern** (Figure 5.6a). The dendritic drainage pattern is usually found in areas that have flat-lying and/or homogenous rock or soil (especially glacial till) underlying the region. Trellis and rectangular drainage patterns are very similar in appearance, but they are the result of different

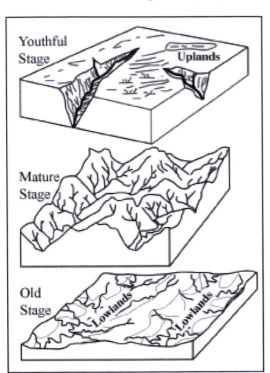

FIGURE 5.5 Landscape Stages.

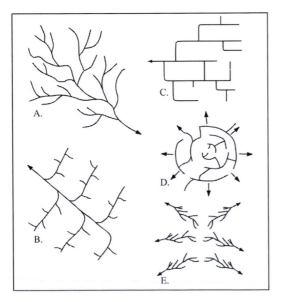

FIGURE 5.6 Stream Drainage Pattern.

processes and location characteristics. The **trellis drainage pattern** (Figure 5.6b) is usually structurally controlled by folded, or thrust-faulted, parallel ridges. The **rectangular drainage pattern** (Figure 5.6c) is fracture-controlled (faults and joints), and commonly occurs in flat-lying sedimentary rock layers. **Annular drainage patterns** (Figure 5.6d) are ring-shaped patterns developed on eroded domes and basins. The rivers of the **radial drainage pattern** (Figure 5.6e) "radiate" downward, in all directions away from mountain peaks, such as volcanoes.

POLLUTION OF RIVERS AND STREAMS

Rivers have traditionally been the "dumping ground" for every kind of waste material humans can produce because the materials are easily swept away and thus disposed of. Before the Industrial Revolution and the great rise in human population, people could drink directly from almost every river, stream, or other body of surface water they came across. Today this is true only for uninhabited regions.

Rivers that have in the past been polluted by sewage, industrial waste, manure from stockyards, and other pollutants are slowly recovering due to the efforts of governments to clean them up. Stricter laws against the dumping of untreated material into waterways have been enacted in response to the unsanitary and otherwise unsafe conditions. One example of the amount of pollution that was common in rivers before environmental laws were passed and enforced is the Cuyahoga River in Ohio. Between the 1930s and the 1960s, flammable materials in the Cuyahoga River caught on fire several times. Since the Clean Water Act was passed in the 1970s, the water quality and appearance of the Cuyahoga have noticeably improved, and plants and animals have repopulated the river system.

EXERCISE 5.1: RIVERS

LAKE MCBRIDE, KANSAS, 7.5-MINUTE SERIES

1. What stream drainage pattern is seen on this map?

2. Locate a dam on this map.

   Where is the dam located?

   What is the name of the dam?

3. What is the morphology of Ladder Creek?

4. Does Ladder Creek have a flood plain? If yes, where is it located (in relation to the creek itself)?

5. In what stage of development is Ladder Creek?

6. In what stage of development is the landscape in the southwest corner of the map?

7. In what stage of development are the creeks in Garvin Canyon, Battle Canyon, and Horsethief Canyon?

8. In what stage of development is the landscape along Timber Canyon?

9. In what direction does the creek in Burris Draw flow? (Hint: look at the contour lines crossing the creek.)

10. What is the relief of Battle Canyon? What is the gradient of Battle Canyon? (Show your work.)

11. What is the relief of Landon Draw? What is the gradient of Landon Draw? (Show your work.)

MANGHAM, LOUISIANA, 15-MINUTE SERIES

1. What are Binion Brake, Clear Lake, and the strangely shaped marsh above Amity Church?

2. Are Big Ridge Slough and Thomason Slough (bottom left) perennial or intermittent streams?

3. What is the morphology of the Boeuf River?

4. At what stage of development is the Boeuf River?

5. At what stage of development is the landscape on this map? (Hint: compare the elevation of the land to the elevation of the river.)

THACKERVILLE, TEXAS, 7.5-MINUTE SERIES

1. What is the morphology of the Red River?

2. At what stage of development is the Red River?

3. Does the Red River have a flood plain?

4. At what stage of development is the stream in the upper-left corner?

5. What is the stream drainage pattern of the stream in the upper-left corner?

6. At what stage of development is the landscape surrounding Bear Head Creek?

7. At what stage of development is Bear Head Creek?

8. What is the relief of Wolf Hollow Creek? What is the gradient of Wolf Hollow Creek? (Show your work.)

ONTARIO, CALIFORNIA, 15-MINUTE SERIES

1. What depositional stream feature is seen at the mouths of the canyons exiting the mountains to the north (San Antonio Canyon, Deer Canyon, Day Canyon, etc.)?

2. At what stage of development is the landscape in the northern portion of the map?

3. What stream drainage pattern is seen in the northern portion of the map?

ELK POINT, SOUTH DAKOTA, 30-MINUTE SERIES

1. What feature of meandering rivers are Lake Goodenough and McCook Lake?

2. At what stage of development is the Missouri River?

3. Does the Missouri River have a flood plain? If yes, where is it located (in relation to the river itself)?

4. What is the morphology of the Sioux River?

5. At what stage of development is the Sioux River?

6. At what stage of development is the landscape in the Big Springs area?

7. At what stage of development is the landscape at Prairie Center?

CORDOVA, ALASKA

1. What is the morphology of the Copper River, south of Miles Lake?

2. What depositional stream feature is found at the mouth of the Copper River?

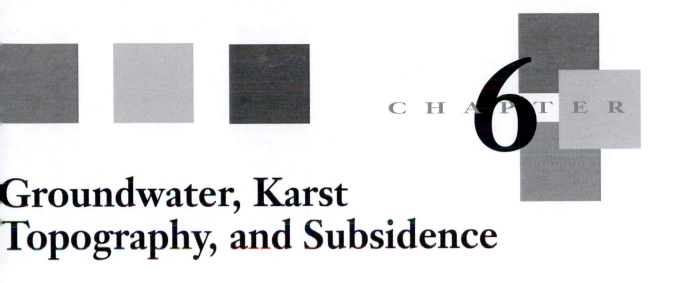

C H A P T E R

6

Groundwater, Karst Topography, and Subsidence

GROUNDWATER

Groundwater is the small percentage of precipitation (rain, snow, etc.) that seeps into the ground, and is an important source of fresh water. The *National Geographic Special Edition: Water*, November 1993, illustrates the percentage of the Earth's water with a gallon jug. With all the Earth's water in the jug, only one tablespoon is fresh water. Ninety-seven percent is salt water, 2% is glacial ice, and the remaining 1% is in rivers, lakes, water vapor in the air, dampness in the soil, and as groundwater.

The depletion of groundwater sources is fast becoming a problem. The groundwater resources in many regions are being depleted faster than the underground storage reservoirs (aquifers) can be naturally filled. The depletion is caused by excessive pumping for municipal water supplies and irrigation. Pumping treated water back into an aquifer is an artificial way of recharging it. Many areas in the state of California practice this type of artificial recharge. Humans are accidentally and/or irresponsibly contaminating groundwater reserves by introducing toxic materials (untreated human and animal wastes, chemicals, occasionally nuclear wastes, and bacteria) into them.

Surface water sites for permanent storage of dangerous or slowly degraded wastes require careful planning, monitoring, and protection from natural disaster, such as flooding. Thorough geologic examination of any proposed disposal site is necessary in order to avoid locating the disposal site on fault lines or other geologically unsuitable areas. Even given great care, this kind of disposal is probably a danger to groundwater supplies because of seepage. Many older waste sites leak because the proper steps were not taken to determine the safety and durability of the sites. Even modern sites must be watched carefully and continuously to guard against leaks. Municipal landfills, another form of waste disposal, are also a source of toxic material that can be introduced into groundwater. All landfills must be sealed with an impermeable layer of clay to prevent the migration (leaching) of poisonous materials such as insecticides, used motor oil, household chemicals, and other contaminants into the groundwater systems.

GROUNDWATER: GENERAL TERMINOLOGY

Several terms are used to describe groundwater and its associated topographic features. **Pores** in rocks, sediments, and soils are the voids (holes) between grains. **Porosity** is a ratio usually expressed as a percent of the total volume of the pore spaces to the volume of the entire volume of the rock, sediment, or soils, including the pore spaces. Fifteen percent

157

is a large porosity for a sandstone, but many soils and sediments have a much larger porosity. **Permeability** is a term used to describe the ease through which water and other fluids, such as air or oil, can pass from pore to pore through a material such as sandstone.

With respect to groundwater, subsurface material is described by the extent to which the pores are filled with water. Unless one is standing in water or on saturated ground, the ground beneath one's feet has pore spaces filled with air enriched by carbon dioxide and perhaps some methane because of organic decomposition. This is the **zone of aeration** or the **vadose zone**. At some depth, which varies with the kinds of rock and the amount of rainfall, there is a **saturated** or **phreatic zone** in which all the pore spaces are completely filled with water. The boundary between the zone of aeration and the phreatic zone is called the **water table**. Despite its name, the water table is not level, but instead, undulates—higher under the hills and lower across and under valleys. Springs and margins (shores) of lakes and permanent rivers or streams mark the position where the land surface penetrates the water table. There is a transitional zone above the water table in the base of the vadose zone where water rises by capillary action (the force resulting from the surface tension of water) to a distance above the water table that depends on the size and shape of the pores. Some of the pores are filled with water.

The water table mimics the ground surface but it usually has a more subdued topography. Because the water table is a three-dimensional surface, it can be contoured, just like the ground surface above. The only requirement is sufficient data points from water wells and other sources to create sufficient control.

The groundwater flows downhill under the influence of gravity and its flow can be represented by flow lines. **Flow lines** intersect the water table contours at right angles. Flow lines cannot intersect with one another, but they can diverge or converge. The groundwater flows until it intersects or enters a stream. In the case of an **influent stream**, one where groundwater flows into the stream, the stream occupies a low area in the water table and pollution will not cross a stream and a pollutant will not enter the groundwater. Groundwater flow is confined to drainage basins like streams. Flows cannot cross major divides but can cross sub-basin divides at the lower end.

AQUIFERS: GENERAL TERMINOLOGY

An **aquifer** is a rock or regolith body that contains, easily absorbs, transmits, and yields groundwater. Poorly cemented *sandstones* and *fractured limestone* make good aquifers because they tend to have good porosity and permeability. The geographic region where meteoric (atmospheric) water enters an aquifer is called the **recharge zone**. When water is prevented from infiltrating into the recharge zone, an aquifer will eventually run dry.

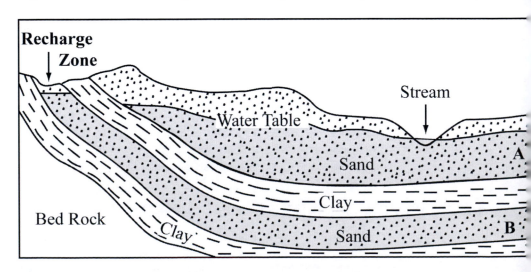

FIGURE 6.1 Water Table Aquifer (A) and Confined Aquifer (B).

As an example, the systematic destruction of many shallow ephemeral (temporary) lakes, buffalo wallows, that once dotted the High Plains increased the acreage that could be devoted to crops and the rate of soil erosion. This hindered waterfowl in a major flyway, and drastically decreased water infiltration. A **confining bed** (layer) is an impermeable (very low permeability) rock or sediment that prevents the movement of fluids. The most common confining beds (layers) are *clays* and *shale*, which have a low porosity. Aquifers with confining beds both above and below the groundwater are called **confined aquifers**. **Artesian flows** occur naturally on the surface in the form of springs or natural water fountains as a result of the high pressure of confined aquifers. The pressure allows the water to flow at ground level naturally without pumping. If the pressure is high enough, a fountain of water will shoot up into the air if the aquifer is penetrated. The fountain will continue to flow as long as the pressure is high enough. Over time, the pressure in the aquifer will drop to a point where it becomes necessary to pump that water to the surface. The most common type of water wells are dug or drilled into groundwater reservoirs and water is recovered by mechanical pumping. Local aquifers, in desert regions, that follow a stream underground are called **perched aquifers**.

The **Ogallala Aquifer** is an extensive aquifer system that underlies the High Plains (South Dakota to Texas) and is a major source of water for irrigation. Groundwater withdrawl, particularly in the Texas Panhandle (Llano Estacado), has exceeded the aquifer's recharge rate. It is hard to imagine that a source of groundwater of one quadrillion gallons could run dry in the future, or become so contaminated that it would eventually become unusable, but it is possible.

The **Gulf Coast Aquifer System** is an example of an extensive groundwater source along the Gulf Coast. The Gulf Coast Aquifer System has been divided into three separate aquifers in Texas, the **Jasper** (lowermost) Aquifer, the **Evangeline** (middle) Aquifer, and the **Chicot** (uppermost) Aquifer. The Chicot and/or parts of the Evangeline aquifers are a supplementary water source for the city of Beaumont, Texas. Saltwater intrusion from the Gulf of Mexico is steadily decreasing the quality of the Gulf Coast Aquifer System because as the fresh water is removed, salt water percolates in and replaces it. The problem of increased saltwater intrusion must be dealt with in the near future.

CAVES AND KARST TOPOGRAPHY

A **cave** is an underground cavity developed by the dissolution of soluble rocks, usually limestone, by groundwater. Water moving through the vadose zone toward the phreatic zone dissolves carbon dioxide from the soil atmosphere and becomes slightly acidic. Most caves develop just below the water table, in the uppermost section of the phreatic zone, where water circulation and acidity is greatest, and where water is least saturated with calcium. **Mammoth Cave** in Kentucky is a large cave system developed by the migration of groundwater throughout a limestone region. A cave that is above the water table (in the vadose zone) will begin a reverse process of filling in with calcite. Such caves are termed "**active**." The atmosphere of the cave is not as rich in carbon dioxide as the soil atmosphere, and may be little different from ordinary air if the cave has openings to the surface. Entering infiltrated water (water dripping from the ceiling into the cave) has dissolved $CaCO_3$ in it, but once in the cave, it loses carbon dioxide, with the result that calcite is precipitated on the walls, ceiling, and floor of the cave. Precipitation from dripping mineralized water produces several distinctive features. **Stalactites** are cave formations that "grow" downward from the ceiling, and **stalagmites** are cave formations that "grow" upward from the cave floor. When stalagmites and stalactites join, they form a **column**. There are many other strange and beautiful cave formations, such as **cave pearls**, **lily pads**, **cave popcorn**, **helictites** (straw-like features), **cave curtains** or **cave draperies**, and **cave onyx**, to name a few. A "dead" cave is one in which cave formations are no longer "growing" because the cave is dry. **Carlsbad Caverns** in New Mexico is, for the most part, a "dead" cave system that developed during the ice age when the region had a more humid climate than today, and when the water table was closer to the surface.

KARST TOPOGRAPHY

The name "karst" is derived from the Karso region of Dalmatia in the Dinaric Alps, southern Croatia, in Eastern Europe. Another very famous karst region is the upland area in the Yunnan and Guizhou provinces, south-central China. The strange vertical topography is often represented in traditional Chinese paintings and drawings.

Karst topography is a descriptive term for regions where the surface landscape has been modified to a large extent by the dissolution of underlying soluble rock, usually *limestone*, but sometimes *gypsum*. The development of karst topography can be more rapid and effective than surface erosion. One of the requirements for the development of karst topography is the presence of soluble rock within the upper part of the water table. The principal component of limestone is **calcite (CaCO$_3$)**, which is quite soluble in even slightly acidic waters. *Dolomite* [Mg,Ca(CO$_3$)$_2$], another rock-forming carbonate mineral, is much less soluble. *Gypsum* is an evaporitic mineral (rock) that is soluble in any water except heavy brine. Another requisite feature of the rock, in addition to solubility, is **permeability**. Limestones often have little **primary permeability** because the calcareous material cements together firmly. The Texas Hill Country, from Georgetown to Kerrville, although not strictly a karst region (there is too much dolomite), has local patches of karstic features. The many fractures that resulted from the movement along the Balcones Fault Zone that borders the eastern margin of the Hill Country has given the limestone a good **secondary permeability** because faulting has fractured the rock.

KARST FEATURES

Landscapes in limestone regions dotted with undrained depressions are karst landscapes. Karst topography is easy to recognize on the landscape and is distinctive on topographic maps. **Sinkholes** are depressions in the Earth's surface that develop when the ceiling of a cave collapses. When enough of the supporting rock in the cave is dissolved, the ceiling will collapse due to gravity. If the cave is "active" and is completely filled with water, the water will help hold up the cave ceiling. Thus, most sinkhole collapses occur when the water table is lowered by drought, long-term climate change, diversion of infiltrating water due to increased urbanization, excessive removal of groundwater by wells, or all of these factors combined. The development of sinkholes on the surface can involve a series of small collapses or one large one. Sinkholes are recognized on a topographic map by circular contour lines that bear hachures. Central Florida is an area with a high water table and an underlying permeable (young) and fairly pure limestone (little dolomite). In Florida, sinkholes often appear abruptly and without warning, to swallow cars, homes, businesses, and sometimes people unlucky enough to be in the area at the time of collapse. If a sinkhole drops beneath the water table, it will become a lake.

When several sinkholes join together they form **solution valleys**, also known as **uvalas**. A solution valley (uvala) is a valley-like depression that forms in a linear fashion along a fault or set of joints. **Disappearing streams** are streams that flow into a sinkhole or disappear underground abruptly. The streams seep into cracks or crevices enlarged by dissolution and continue to flow underground through cave systems. Some disappearing streams reappear further downslope if the water table is higher than the slope of the topography. Most karst features that develop on land can also be found offshore. The movement of groundwater can form freshwater caves below the ocean floor. If the cave ceiling collapses, freshwater sinkholes will develop beneath the saltwater ocean. The fresh water and salt water will mix together near the top of the sinkhole, but some fresh water can be found near the bottom of the sinkhole.

Karst Topography Stages ("Ages") of Development

The topographic features associated with karstic areas indicate the relative amount of time the area has been exposed to the processes of dissolution. The stages of karst topography, like stream stages, are described in relative order.

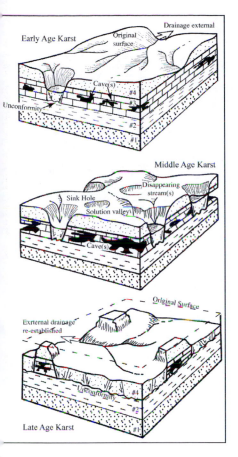

FIGURE 6.2 Karst Topography.

The **pre-karst stage** is characterized by the presence of surface streams, an external drainage, and little or no subsidence evident in the region. Pre-karst topography is hard to recognize on a topographic map, unless the observer is familiar with the existence of soluble rock underground and the presence of groundwater.

Early-age karst topography (Figure 6.2) is characterized by the appearance of a few sinkholes and disappearing streams, but most of the streams exit the area.

Middle-age karst topography (Figure 6.2) is characterized by the presence of numerous sinkholes, disappearing streams, solution valleys (uvalas), and an extensive internal drainage system. Few streams are able to exit the area. When the groundwater is near the surface and sinkholes form, sinkhole lakes are common.

In **late-age karst topography** (Figure 6.2), the dissolution of the underlying rock is mostly completed. The region is redeveloping an external drainage. Very little of the original limestone is left, and an unconformable surface between the rock originally overlying the limestone and the rock originally underlying the limestone is formed. A few **monadnocks** (erosional remnants) remain supported by limestone that survived dissolution.

Subsidence

Landscapes change continuously due to natural processes of erosion, tectonics, plants and animals, and by human intervention. Land will subside or submerge (sink) in areas due to a variety of causes. Natural causes include the dissolution of underlying rock, lowering of groundwater levels due to climatic changes, tectonic processes, and by compaction (collapse of pore spaces) due to the weight of the overburden. Mud and clay often have an original porosity that approaches 80%. The area in and around the Mississippi Delta is subsiding under its own weight because of the compaction of the river-deposited mud. When groundwater is withdrawn at a rate greater than the recharge rate, then local subsidence may occur. Subsidence is also caused by the withdrawal of petroleum, natural gas, sulfur, and salt. The removal of oil, gas, and sulfur from the Spindletop Dome, Beaumont, Texas, has caused enough local subsidence to create a shallow lake (approximately 10 feet deep) in an area that was originally the highest in the area. Parts of the NASA complex, Pasadena, Texas, and areas in Baytown, Texas, are currently subsiding at rates approaching 2 cm/yr as a result of compaction of underlying sediment enhanced by the excessive withdrawal of groundwater by wells. This is producing a local transgression, and areas once dry, yet not very high, are now under marine water. Natural subsidence cannot be avoided or controlled. One solution for man-made subsidence is to pump treated water back into aquifer systems.

EXERCISE 6.1

Answer the following questions.

LAKE WALES, FLORIDA, MAP

1. What feature of karst topography are Lake Wales, Lake Mabel, and Lake Starr?

2. How did Lake Wales, Lake Mabel, and Lake Starr form?

3. In general, what are circular contour lines with hachure marks?

4. What feature of karst topography is represented by circular contour lines with hachure marks?

5. At what stage of development is the landscape in this area?

6. What evidence supports the previous answer?

7. What is the elevation of Lake Starr?

8. What kind of rock is most likely to be found in this area (Florida)?

9. What type of stream might you find, if present, associated with karst topography?

0. Why is Lake Belle so strangely shaped?

GENERAL

1. Define confined aquifer.

2. Can an aquifer run out of accessible water? If yes, how or why can this happen?

3. Which rocks make good aquifers?

4. Which rocks/sediments make good confining layers (beds)?

5. Which rocks are prone to dissolution by acidic water?

6. What karstic feature forms when water dissolves rock below ground?

7. What is the chemical composition of most rocks that undergo dissolution?

8. List one famous cave system.

9. What processes contribute to or directly cause subsidence?

10. Is it a good idea to build a home or business in a known karstic area without a subsurface survey? Explain your answer.

EXERCISE 6.2: GROUNDWATER FLOW

Draw the flow lines on the following map from points A, B, and C. Remember, flow lines are always intersect the contour lines at right angles.

From which point, A, B, or C, will the groundwater intersect the stream at the highest elevation?

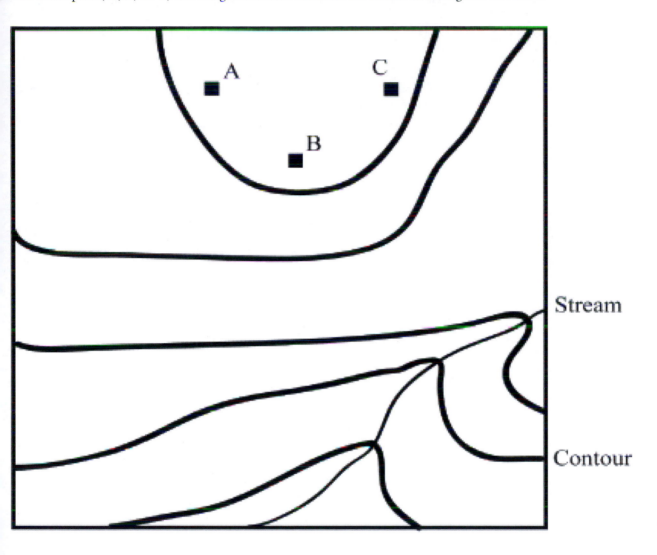

EXERCISE 6.3: GROUNDWATER PROBLEM

The following map is a hypothetical groundwater map. A property owner living at point B intends to drill a well to supply water to his home. If the ground surface elevation at point B is 760 feet, what is the minimum depth the well will have to be to reach the water table?

If a toxic substance is accidentally added to the groundwater at point C, will the water at point B be contaminated? Explain your answers.

Will the water at point A be contaminated? Explain your answer.

Will the water at point D be contaminated? Explain your answer.

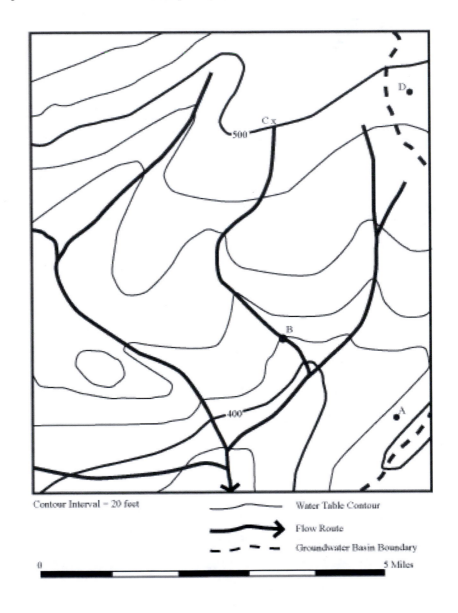

Shorelines

A **shoreline** is the boundary where land and water meet. Shorelines are either classified as **emergent** when seas withdraw and dry land gradually appears, or as **submergent** when the shoreline is (or has recently been) advancing inland. **Emergent shorelines** are the result of regression, the lowering of sea level due to the uplift of the land or by a worldwide change (lowering) of the sea level. **Submergent shorelines** are the result of transgression, the rising of sea level due to the subsidence of the land or by a worldwide change (raising) of the sea level. Broadly speaking, emergent shorelines have fewer inlets and bays than submergent shorelines. Shorelines in areas recently heavily affected by valley glaciation (fjord shorelines) are extensively digitated (fingerlike) even though they can be either emergent or submergent.

GENERAL SHORELINE FEATURES

Much of the sediment in the seas originates on the land. Rivers erode the land and transport the sediment to the sea. The **longshore current** is a current of water that moves parallel to the shore and transports the eroded sediments down-current (in the direction of the prevailing wind, or away from a storm center) and is responsible for the formation of many shoreline features (Figure 7.1).

Bays are recesses in the shoreline that form when materials of differing hardness are eroded away at different rates. The less-hard material is eroded away faster, leaving harder rocky protrusions called

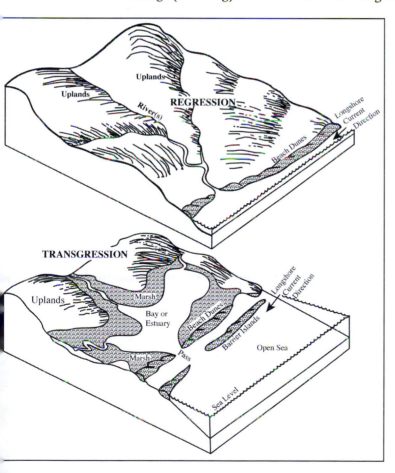

FIGURE 7.1 Shorelines.

167

headlands. The headlands, over time, may become completely cut off from the shore and form **marine arches**, **blowholes**, or small rocky islands known as **stacks**. A **spit** is a sand bar attached to the mainland formed by the deposition of sediments moved by the long shore current. Spits elongate in the direction of the current flow. Overtime, a spit may grow completely across the mouth of a bay and is referred to as a **baymouth bar**. The forme bay, now a **lagoon**, has been separated from the ocean, except for a shallow **tidal inlet** o **pass** (a small opening between the lagoon and the ocean). Lagoons also develop wher barrier islands form along coastlines, such as Laguna Madre along the Texas Gulf Coast **Barrier islands** are elongated sandbars that parallel the coastline and form a "barrier" be tween the open ocean and the coastline. **Tied islands** are islands that act as breakwater to slow the longshore current, and consequently are attached to the coast by sediments de posited by the current. The sandbar that "ties" the island to the shore is called a **tombolo** Tied islands generally form in coastal environments with low amounts of wind.

The shoreline features found on modern coasts were formed in the geologic past and can be recognized in the stratigraphic record. Recognition of these ancient features in sub surface sedimentary rock allows the geologist to pinpoint areas most likely to be source of oil, gas, and fresh water. Some shorelines are rocky and have steep promontories (head lands), and some have masses of sediment (usually sand) called **beaches**. Rocky shore lines, generally in areas of rugged topography and where water depth increases rapidl offshore, change very slowly because rocks resist erosion. Rocky and steep shorelines occu in areas of erosion, whereas beaches represent areas of deposition. For beaches, positio (relative to some fixed referent such as a beach highway), appearance, and shape can shif subtly on a daily basis. Weather and tidal variations influence wave action and sedimen transportation, but most beaches show little if any net changes over periods of years. I might seem like violent storms, such as hurricanes, which commonly cause extensiv beach erosion and change the beach substantially in a matter of hours, might provide ar exception to the preceding statement. But such changes, even though they may remove th beach highway ("fixed referents"), within a few months, will be unnoticeable, as the beacl will look very much as it did before the storm. Of course, some fundamental change in th frequency of violent storms over a long period of time would produce net changes in th general position of the beach and the shoreline by changing the **wave energy** (how har and how frequently waves hit the shore) and the force and extent of the longshore current: Longshore currents control the transfer (migration) and deposition of sediments alon shorelines to an even greater extent than the tides do in many areas.

Beach drifting is the down-current movement of sand accompanying the longshor current. Wave fronts are bent to become more nearly parallel to the shore (wave-travel d rection is bent to become more nearly perpendicular to the beach) as the waves near th shore and where the water shallows to about one-half the wavelength. However, the benc ing, called **refraction**, is rarely complete, with the result that the wave energy strikes th beach at a slight angle to the backwash, and water and sediment are forced to slip sideway in a given direction along the beach.

Conditions of emergence and submergence depend on the balance between sets of of erations (Figure 7.2). Many are commonly working at the same time, and some would, b themselves, produce emergence; others may be working in the opposite "direction" (sut mergence). The following are some of the possible factors, arranged in order of the scop of the effects.

SEA-LEVEL CHANGES: EUSTATIC, LOCAL, AND REGIONAL

Worldwide Eustatic Sea Level(s)

Eustatic sea-level change is a change in the volume of water in the oceans and seas, or i the volume of the ocean basins, that affects shorelines worldwide, at the same tim Historical records and geological evidence show that sea level changes as the volume of ic stored in continental ice caps and glaciers (particularly in Antarctica and Greenland) varie with changes in the global climate. Sea level is rising at present. Plate tectonics, whic

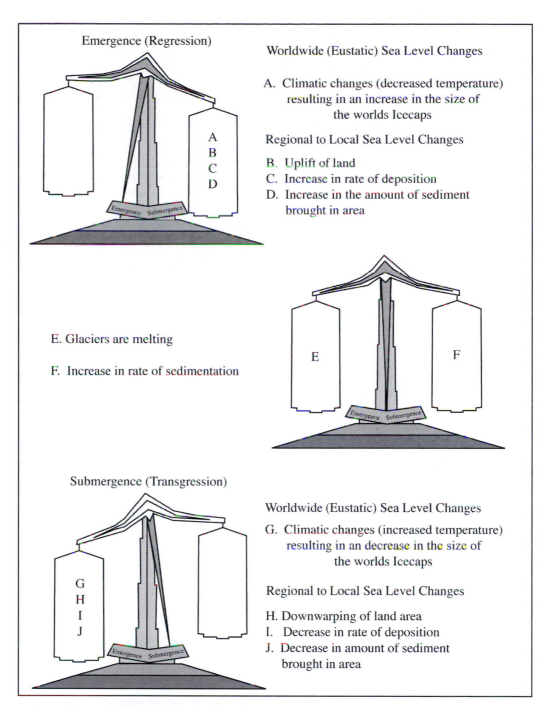

FIGURE 7.2 Causes of Emergence, Equilibrium and Submergence.

affects the motion of lithospheric plates, changes the shape of the ocean basins, and rates of spreading change heights and volumes of oceanic ridges, resulting in eustatic changes.

Regional and Local Effects of Sea-Level Changes

Tectonic activity influences the Earth's surface regionally or locally. There can be uplift in tectonically active areas (compressional and transform margins), downwarping or rebound (when downwarping stops) in back-arc basins, and downwarping induced by crustal cooling at locked plate margins. One of the things that makes many features of our planet hard to measure, including what is exact sea level, is that few if any parts of the Earth's crust "hold still."

Climatic factors influence the rate of weathering and erosion, and thus the production of sediment varies. Storms on land create floods that move sediment to the shoreline (eventually), and storms at sea drive the longshore currents and determine wave size. Storminess affects the transfer or deposition of sediments along the shore, and the rate at which promontories (headlands) are attacked. The "greenhouse effect" and possible global warming may be responsible for the eustatic rise in sea level (transgression) that is currently submerging shorelines and encouraging beach erosion.

Human activity also influences the Earth's surface regionally or locally. Clear-cutting a forest may affect local climate, and frees a great deal of sediment because deforestation encourages runoff and erosion. Damming a river temporarily prevents sediment from reaching the shoreline, so the beaches and marshes, deprived of sediment replenishment, erode. The building of **jetties** (usually a pair of structures built along the shoreline to protect a entrance to a harbor, an inlet, or at the mouth of a river) and **groins** (structures built perpendicular to the shoreline to trap sand and prevent erosion) along the shore will cause the longshore current to deposit sediment immediately up-current of the obstruction, and to erode beaches further down-current.

EMERGENT SHORELINES: CAUSES AND CHARACTERISTICS

Emergence of a shoreline (regression) can take place for fundamentally different reasons. One is *actual uplift of the land* relative to sea level. A second is *eustatic sea-level fall*, and the third reason is an *increase in the rate of deposition* (usually a local effect). The amount of sediment deposited exceeds the capacity of the agents of erosion (wave energy and longshore currents) to carry it away. Although it is useful for purposes of analysis to think of these factors separately, they more usually work in combination.

Features of Emergence by Actual Uplift

Wave-cut platforms develop when, in the course of long-term emergence, there is a pause, a temporary reversal, producing brief erosional submergence. Waves plane off (or build up) a smooth surface that slopes gently seaward. **Marine terraces** are wave-cut platforms that are exposed when sea level drops (or land rises). Beaches, lagoons, or backbeach marshes produce a **ridge and swale topography** (**strand plains** and **chenier plains**) that can be left behind when the sea regresses.

Features of Emergence by Deposition

Deltas are, in map view, triangular, fan-shaped, or "bird-foot"-like (Mississippi River Delta) deposits of river sediment, which the stream transports to and deposits in a relatively still body of water, such as a lake or ocean. The river slows (capacity decreases rapidly) and drops its load of sediment in a shape that depends on both the strength of the waves and currents in the body of "quiet" water, and on the texture of the sediment. The river divides into **distributaries** (rivers that have more than one mouth). As deltas grow, they get flatter and flatter, and with time the river migrates laterally rather abruptly (**avulsion**). A delta is a complex of smaller deltas of different, cross-cutting ages. Emergent shorelines are generally straight and smooth, but in areas where the currents and waves are weak, or the amount of sediment introduced by the rivers is very large, deltas will have bumps on, or even complicated projections from, an uncomplicated coast. Along the Gulf Coast, the Mississippi Delta is an example of a situation where enormously more sediment is supplied than the longshore currents can carry away. The delta of the Brazos River is much less impressive because the river supplies barely more sediment than the longshore currents can remove.

Barrier islands (Figure 7.1) are elongated sandbars that parallel the coast with beaches and, commonly, dune ridges. Barrier islands along the Atlantic Coast of the United States from Long Island to northern Florida are discontinuous, and tidal passes (inlets) are important in their development. Texas barrier islands along the Gulf Coast (Galveston

South Texas and Tamaulipas), where tides are small, have few natural passes and are much more effective at cutting off **lagoons** (such as Laguna Madre) from the open sea. Barrier islands along the Texas coast began to develop less than 5,000 years ago, when eustatic sea-level rise (from the melting of the glaciers of the last ice age) slowed, and the rate of deposition began to exceed the rate of drowning of the paleoriver valleys. Texas "bays" or **estuaries** are submerged river valleys.

SUBMERGENT SHORELINES: CAUSES AND CHARACTERISTICS

The causes of submergence of shorelines are, with one exception, the opposites of the causes of emergence. Submerged shorelines (Figure 7.1) can occur via the "drowning" of a coast due to the eustatic rise of sea level (worldwide effect), tectonic downwarping (regional effect), and/or the compaction of thick sediment (along the Gulf Coast, for instance), also a regional effect. Locally, withdrawal of large amounts of water, oil, or natural gas from unconsolidated sediments can speed up subsidence dramatically. Another cause of submergent shorelines is erosion. The forces of erosion can be increased by climatic change or increased storminess, or by the decrease in the amount of sediment brought in, so that the capacity of erosional agents exceeds the sediment supply.

Features of Submergence by Actual Sea-Level Rise

The main identifying feature of drowned submergent shorelines is the presence of clusters of islands (if the topography was hilly), bays, and estuaries. A **bay** (Figure 7.1) is a broad recess in the shoreline. Small, narrow bays not dominated by a river may be called **inlets,** and **fjords** (drowned U-shaped valleys) are long narrow inlets cut by a glacier. An **estuary** is a broadened and tide-dominated, drowned lower portion of a river valley, with brackish water (a mix of sea and fresh water—**hyposaline**). There is little difference between high and low tide along the Texas shoreline, and many rivers in South Texas do not deliver enough fresh water to keep the "bays" from becoming **hypersaline** (saltier than sea water) in the summer due to evaporation. But in many respects, Texas "bays" (such as Sabine Lake) would qualify as estuaries. Chesapeake Bay, the drowned confluence of the Susquehanna and Potomac rivers (Maryland and Virginia), is a classic estuary.

As the sea level rises or the land subsides, these shoreline recesses migrate further and further inland. Bays, particularly large bays, usually have beaches, but low wave energy and poorly developed beaches are characteristic of estuaries. Bays and estuaries usually have marshy or swampy edges. As the sea advances into one of these kinds of recesses, the beach or marsh moves inland. When the water moves inland against a steep or rocky coast, it may produce **wave-cut cliffs**, which, if erosion dominates, temporarily will have **wave-cut platforms**, and perhaps narrow beaches at their bases. **Stacks** (as mentioned at the beginning of the chapter), such as off the California coast, are small, steep-sided rocky islands produced when wave action erodes through a **promontory** or **headland** along a steep, rocky shore.

Features of Submergence by Erosion

Submergence by erosion has the effect of simplifying, straightening, or rounding the shoreline, and making it steeper. When a river avulses (rapidly changes course), abandoning an old delta to build a new one, the old delta, both because it no longer receives sediment and because sediment already deposited continues to compact under its own weight, is rapidly eroded and submerged. Slow eustatic rise in sea level or tectonic downwarping, if unaccompanied by sufficient increase in sediment transfer, will produce the same effect on a regional basis (and, of course, eustatic rise in sea level affects shorelines globally). Regionally, differences in the durability of the materials being eroded may lead to complication in the shape of the shoreline, but nevertheless, where erosion is dominant, simplification is the general rule. Usually in areas of regional submergence, erosion does round or straighten shorelines in promontory or headland areas.

EXERCISE 7.1 SHORELINES

Answer the following questions using the topographic maps provided.

KINGSTON, RHODE ISLAND, 7.5-MINUTE SERIES

1. What shoreline feature is Green Hill Beach?

2. What feature of shorelines is Green Hill Pond?

3. What is Green Hill Beach made of?

4. Where does the source material for beaches originate?

5. How was the beach material transported to the beach?

6. How was the beach material transported along the shoreline?

PROVINCETOWN, MASSACHUSETTS, 7.5-MINUTE SERIES

1. What feature of shorelines is Long Point?

2. In what direction does the longshore current flow on the northern side of the area? (Hint: Look at the shape and rounding of the shoreline.)

3. In what direction is the longshore current flowing on the western side of the area?

4. What feature of shorelines is Pilgrim Lake?

5. Are there any estuaries on this map?

6. List the name of one bay on this map.

CORDOVA, ALASKA

1. What feature of shorelines are Egg Islands, Copper Sands, and Strawberry Reef?

2. What are Egg Islands, Copper Sands, and Strawberry Reef made of?

3. In what direction is the longshore current flowing in the Gulf of Alaska?

4. What feature(s) of shorelines did you use to determine the direction of the longshore current?

5. List the name of one fjord on this map.

6. What type of river is the Copper River, south of Miles Lake? (Refer to Chapter 5.)

7. What depositional stream feature is found at the mouth of the Copper River?

8. What are the white areas on this map?

BLAKELY ISLAND, WASHINGTON, 7.5-MINUTE SERIES

1. What feature of shorelines is the island between Shoal Bay and Humphrey Head?

2. What feature of shorelines connects the previous feature to the mainland?

3. What feature of shorelines are Diamond Point (northwest corner) and Fauntleroy Point?

4. What feature of shorelines borders the eastern side of Blakely Island?

5. What feature of shorelines are Black Rock, Spindle Rock, and Brown Rock?

6. What feature of shorelines is Spencer Spit?

7. What is Spencer Spit made of?

REFERENCES

Allen, John R. L. *Sedimentary Structures Their Character and Physical Basis*. Amsterdam: Elsevier, 1984.

Anders, R. B., McAdoo, G. D., and Alexander, W. H., Jr. "Ground-Water Resources of Liberty County, Texas." Texas Water Development Board Report 72, 140 pp., 1968.

Army Corp of Engineers. "The Mississippi River and Tributaries Project." Available at http://www.mvn.usace.army.mil/pao/bro/misstrib.htm.

Atwater, Brian F., Cisternas V., Marco, Bourgeois, Joanne, Dudley, Walter C., Hendley II, James W., and Peter H. Stauffer. "Surviving a Tsunami—Lessons from Chile, Hawaii, and Japan." United States Geological Survey Circular 1187, 1999. Available at http://pubs.usgs.gov/circ/c1187/#debris.

Baker, E. T., Jr. "Geology and Ground-Water Resources of Hardin County, Texas." Texas Water Commission Bulletin 6406, 179 pp., 1964.

Baker, E. T., Jr. "Hydrology of the Jasper Aquifer in the Southeast Texas Coastal Plain." Texas Water Development Board Report 295, 64 pp., 1986.

Bell, P., and Wright, D. *Rocks and Minerals*. New York: Macmillan Publishing Company, 1985.

Blatt, H., Tracey, R. J., and Owens, B. E. *Petrology: Igneous, Sedimentary, and Metamorphic*. New York: W. H. Freeman and Company, 2006.

Chernikoff, Stanley. *Geology: An Introduction to Physical Geology*. New York: Worth Publishers, Inc., 1995.

Cooper, R. W. "Physical Geology Lecture Outline." Lamar University, Department of Geology. Unpublished.

De Castella, Tom. "The Eruption That Changed Iceland Forever," *BBC News Magazine*, http://news.bbc.co.uk/go/pr/fr/-/2/hi/uk_news/magazine/8624791.stm, April 16, 2010.

Encyclopedia Britannica, S.v. "Mount Saint Helens."

Encyclopedia Britannica, S.v. "Crater Lake."

Encyclopedia Britannica Annual, 1990, p. 170.

Federal Emergency Management Agency. Available at http://www.fema.gov/pdf/hazards/earthquakes/nehrp/fema-253_unit3b.pdf.

Foster, Robert J. *Geology*. Columbus: Charles E. Merrill Publishing Company, 1980.

Kosar, Kevin R. "Disaster Response and Appointment of a Recovery Czar: The Executive Branch's Response to the Flood of 1927." Congressional Research Service, The Library of Congress, October 25, 2005, p. 11. Available at http://www.fas.org/sgp/crs/misc/RL33126.pdf.

Levin, Harold L. *The Earth Through Time*. United States of America: Saunders College Publishing, 1999, p. 163.

Mairson, Alan. "The Great Flood of '93," *National Geographic*, Vol. 185, No. 1, January 1994, pp. 42–81.

Monroe, J. S., and Wicander, R. *Physical Geology, Exploring the Earth*. Minneapolis: West Publishing Company, 1995.

Montgomery, Carla W. *Physical Geology*. Dubuque: Wm. C. Brown Publishers, 1993.

Mottana, A., Crepsi, R., and Liborio, G. *Simon and Schuster's Guide to Rocks and Minerals*. New York: Simon & Schuster Inc., 1978.

Parfit, Michael. "Sharing the Wealth of Water," in Water: The Power, Promise and Turmoil of North America's Fresh Water, *National Geographic Special Edition*, November 1993, pp. 20–36.

Pirsson, Louis V. *Rocks and Rock Minerals*. New York: John Wiley & Sons Inc., 1906.

Schumn, Stanley A. *The Fluvial System*. New York: John Wiley & Sons Inc., 1977.

Tennissen, A. C. *Nature of Earth Materials*. New Jersey: Prentice-Hall, Inc., 1974.

Wesselman, J. B. "Geology and Ground-Water Resources of Orange County, Texas." Texas Water Commission Bulletin 6516, 112 pp., 1965.

Wesselman, J. B. "Ground-Water Resources of Chambers and Jefferson Counties, Texas." Texas Water Development Board Report 59, 136 pp., 1971.

Wicander, Reed, and Monroe, J. S. *Essentials of Geology*, 3rd edition. United States of America: Thomson Learning, 2002.

Woods, K. M. "The Gulf Coast Aquifer System in Southeast Texas: A Study of Groundwater Resources," in Cooper, R. W., et al. *Transect of the Upper Gulf Coast: Geology, Resources, Environment*. Lamar University, Department of Geology, 1996.

Zwingle, Erla. "Wellspring of the High Plains," *National Geographic*, Vol. 183, No. 3, March 1993, pp. 81–109.